Zamir K. Punja, PhD
Editor

Fungal Di...
in...
Biochemistry,
and Gene...

Pre-publication
REVIEWS,
COMMENTARIES,
EVALUATIONS . . .

"**I** have been looking for a book like this for a while! The expert authors give an excellent synthesis of a complex discipline focusing on the key areas of signal transduction, cell biology, hypersensitive response, resistance genes/ proteins, and progress in genetic engineering to enhance disease resistance."

John Harper, PhD
Lecturer in Plant Science
and Microbiology,
Shool of Agricultural
and Veterinary Sciences,
Charles Sturt University,
Australia

"**A**s one of the leading scientists in this rapidly evolving field of research, the editor has made an exceptionally representative selection of renowned contributors for this book. This book addresses the most recent advances in the study of fungal disease resistance and illustrates the successful application of molecular, biochemical, cellular, and genomic approaches. Each chapter is well written and easy to follow. This book will be invaluable to researchers as well as educators. For the former, it covers the most up-to-date advances, and for the latter it will be an indispensable reference for lectures."

Tim Xing, PhD
Assistant Professor,
Department of Biology
and Institute of Biochemistry,
Carleton University, Ottawa

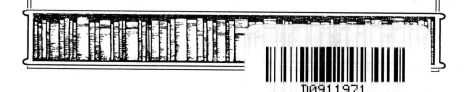

More pre-publication
REVIEWS, COMMENTARIES, EVALUATIONS . . .

Food Products Press®
An Imprint of The Haworth Press, Inc.
New York • London • Oxford

Fungal Disease Resistance in Plants

Biochemistry, Molecular Biology, and Genetic Engineering

FOOD PRODUCTS PRESS®
Crop Science
Amarjit S. Basra, PhD
Senior Editor

In Vitro Plant Breeding by Acram Taji, Prakash P. Kumar, and Prakash Lakshmanan

Crop Improvement: Challenges in the Twenty-First Century edited by Manjit S. Kang

Barley Science: Recent Advances from Molecular Biology to Agronomy of Yield and Quality edited by Gustavo A. Slafer, José Luis Molina-Cano, Roxana Savin, José Luis Araus, and Ignacio Romagosa

Tillage for Sustainable Cropping by P. R. Gajri, V. K. Arora, and S. S. Prihar

Bacterial Disease Resistance in Plants: Molecular Biology and Biotechnological Applications by P. Vidhyasekaran

Handbook of Formulas and Software for Plant Geneticists and Breeders edited by Manjit S. Kang

Postharvest Oxidative Stress in Horticultural Crops edited by D. M. Hodges

Encyclopedic Dictionary of Plant Breeding and Related Subjects by Rolf H. G. Schlegel

Handbook of Processes and Modeling in the Soil-Plant System edited by D. K. Benbi and R. Nieder

The Lowland Maya Area: Three Millennia at the Human-Wildland Interface edited by A. Gómez-Pompa, M. F. Allen, S. Fedick, and J. J. Jiménez-Osornio

Biodiversity and Pest Management in Agroecosystems, Second Edition by Miguel A. Altieri and Clara I. Nicholls

Plant-Derived Antimycotics: Current Trends and Future Prospects edited by Mahendra Rai and Donatella Mares

Concise Encyclopedia of Temperate Tree Fruit edited by Tara Auxt Baugher and Suman Singha

Landscape Agroecology by Paul A Wojkowski

Concise Encyclopedia of Plant Pathology by P. Vidhyasekaran

Molecular Genetics and Breeding of Forest Trees edited by Sandeep Kumar and Matthias Fladung

Testing of Genetically Modified Organisms in Foods edited by Farid E. Ahmed

Fungal Disease Resistance in Plants: Biochemistry, Molecular Biology, and Genetic Engineering edited by Zamir K. Punja.

Plant Functional Genomics edited by Dario Leister

Immunology in Plant Health and Its Impact on Food Safety by P. Narayanasamy

Abiotic Stresses: Plant Resistance Through Breeding and Molecular Approaches edited by M. Ashraf and P.J.C. Harris

Multinational Agribusinesses edited by Ruth Rama

Sustainable Plant Disease Management by Jagtar S. Dhiman

Teaching in the Sciences: Learner-Centered Approaches edited by Catherine McLoughlin and Acram Taji

Durum Wheat Breeding: Current Approaches and Future Strategies edited by Conxita Royo, M. M. Nachit, N. Di Fonzo, J. L. Araus, W. H. Pfeiffer, and G. A. Slafer

Fungal Disease Resistance in Plants

Biochemistry, Molecular Biology, and Genetic Engineering

Zamir K. Punja, PhD
Editor

Food Products Press®
An Imprint of The Haworth Press, Inc.
New York • London • Oxford

Published by

Food Products Press®, an imprint of The Haworth Press, Inc., 10 Alice Street, Binghamton, NY 13904-1580.

Cover design by Lora Wiggins.

Cover images by Zamir K. Punja, PhD.

Library of Congress Cataloging-in-Publication Data

Fungal disease resistance in plants : biochemistry, molecular biology, and genetic engineering / Zamir K. Punja, editor.
    p. cm.
  Includes bibliographical references and index.
  ISBN 1-56022-960-8 (alk. paper)—ISBN 1-56022-961-6 (pbk. : alk. paper)
  1. Fungal diseases of plants. 2. Plants—Disease and pest resistance. 3. Plant-pathogen relationships. 4. Plant molecular biology. 5. Plant genetic engineering. I. Punja, Zamir K.
  SB733.F86 2004
  632'.4—dc22

                                       2003024391

CONTENTS

Chapter 5. Pathogenesis-Related Proteins and Their Roles in Resistance to Fungal Pathogens   139

Jayaraman Jayaraj
Ajith Anand
Subbaratnam Muthukrishnan

Chapter 6. Induced Plant Resistance to Fungal Pathogens: Mechanisms and Practical Applications   179

Ray Hammerschmidt

Chapter 7. Genetic Engineering of Plants to Enhance Resistance to Fungal Pathogens   207

Zamir K. Punja

ABOUT THE EDITOR

Zamir K. Punja, PhD, is Professor in the Department of Biological Sciences at Simon Fraser University in Burnaby, British Columbia. He previously worked for the Campbell Soup Company as a research scientist working on vegetable pathology/biotechnology at various research facilities. He has published over 110 peer-reviewed journal articles, six review papers, and twelve book chapters covering the areas of plant pathology, plant tissue culture, and genetic engineering of plants for disease resistance. He is currently the Editor-in-Chief of the *Canadian Journal of Plant Pathology.*

Dr. Punja's research interests include the development of methods to manage fungal diseases on vegetable crops, with emphasis on the applications of genetic engineering and biological control methods. He served as President of the Canadian Phytopathological Society (CPS) and is a member of numerous professional societies, including the American and Canadian Phytopathological Societies, the British Mycological Society, the American Society of Plant Biologists, the Mycological Society of America, and the International Tissue Culture Association.

CONTRIBUTORS

Ajith Anand received a BSc from Calicut University and an MS in botany from Maharaja Sayajirao University, Baroda, Gujarat, India. He received a PhD in botany from Madras University, India. He served as a postdoctoral fellow in the Department of Biochemistry, Kansas State University, during 1999-2003. He is currently a research associate working in the Plant Biology Division at the Samuel Noble Foundation, Oklahoma. His research interests include purification and characterization of pathogenesis-related proteins, transgenic approaches for enhancing disease resistance, assaying the antimicrobial/antifungal activity of PR proteins, systemic disease signaling in plants, and structural and functional genomics of plant genes involved in *Agrobacterium* plant transformation.

P. J. G. M. de Wit studied at Wageningen University, the Netherlands, where he obtained an MSc and a PhD in biology. He spent one year as a postdoctoral fellow in the Department of Plant Pathology, University of Kentucky, Lexington. He returned to Wageningen and became a full professor of phytopathology at Wageningen University in 1990. His research interests are in molecular plant pathology and molecular breeding. He and his research team were the first to isolate a fungal avirulence gene (*Avr9* of *Cladosporium fulvum*) in 1991. He is presently head of the Laboratory of Phytopathology and director of the graduate school in Experimental Plant Sciences.

Ray Hammerschmidt obtained a BSc in biochemistry and an MSc in plant pathology from Purdue University, Indiana. He received a PhD in plant pathology from the University of Kentucky, Lexington. In 1980, he joined the faculty at Michigan State University as an assistant professor of plant pathology. He is currently Professor and Chairperson of the Department of Plant Pathology at Michigan State University. His research interests are in the induced defenses of plants to diseases, the role of phytoalexins and other biochemical compounds in disease resistance, and the applications of induced resistance against fungal pathogens.

Michèle C. Heath received a BSc and PhD from the University of London. She has spent most of her academic career at the University of Toronto, Canada, where she is currently an emeritus professor; she is also an adjunct professor at the University of Victoria. She has been a president of the Canadian Phytopathological Society, the editor-in-chief of the journal *Physiological and Molecular Plant Pathology*, and is a fellow of the Royal Society of Canada. Her research interests focus on the cellular and molecular responses of plants to infection by biotrophic fungi, the role of nonhost resistance to disease, the hypersensitive response in plants, and programmed cell death in response to infection by fungi.

Jayaraman Jayaraj received a BSc in agriculture, and an MSc and a PhD in plant pathology from Annamalai University, India. He served as a senior lecturer of plant pathology at Annamalai University from 1991 to 1999. He joined Kansas State University in 1999 and is currently an associate scientist. His research interests include biological control of plant pathogens, transformation of plants with PR proteins and other genes of agronomic importance, purification of PR proteins, expression of pharmaceutical proteins in plants, and induced systemic resistance in plants.

M. H. A. J. Joosten studied horticulture at Wageningen University, the Netherlands, and received an MSc in 1986. He obtained a PhD in 1991 from the Department of Phytopathology, in the group of Pierre J.G.M. de Wit at Wageningen University. After a few postdoctoral positions in phythopathology, during which he isolated the *Avr4* gene of *Cladosporium fulvum,* he obtained a position as assistant professor in this department. He is currently co-supervisor of the *Cladosporium* group, together with Pierre J.G.M. de Wit. His research interests focus on signal perception and defense signaling in plants showing gene-for-gene resistance against pathogens.

Erich Kombrink studied chemistry and biochemistry and obtained a PhD in biochemistry at Philipps University in Marburg, Germany, with emphasis on enzymology and regulation of starch and carbohydrate metabolism in plants. He continued this work as a postdoctoral fellow at the University of California, Santa Cruz, in association with Dr. Harry Beevers. In 1983, he joined the Max-Planck Institute for Plant Breeding Research in Cologne, Germany, where he worked in the Biochemistry Department with Dr. Klaus Hahlbrock as group leader and senior scientist on various aspects of plant defense mechanisms against pathogens. His research interests include PR proteins, phenylpropanoid metabolism, gene regulation, and structure-function relationships of enzymes of phenylpropanoid metabolism. Recently, he has been involved in research in the Department of Plant-Microbe Interactions, headed by Dr. Paul Schulze-Lefert, studying protein profiling and proteomics in plant-pathogen systems.

Denny G. Mellersh received a BSc and PhD from the University of Toronto, Canada, where he studied the expression of plant resistance and susceptibility to biotrophic fungal pathogens. He is currently a postdoctoral fellow in the Sainsbury Laboratory at the John Innes Centre in the United Kingdom, where his current research interests focus on how pathogenic and symbiotic biotrophic fungi establish compatibility with susceptible host plants, and on the cellular and molecular changes in plant cells that accompany or accommodate biotrophic infection.

Subbaratnam Muthukrishnan received a BSc and an MSc in biochemistry from the University of Madras, India, and a PhD in biochemistry from the Indian Institute of Science. He was a postdoctoral scientist at the University of Chicago (1971-1973), the Roche Institute of Molecular Biology (1973-1976), and the National Institutes of Health (1976-1980). He joined Kansas State University in 1980, where he is currently Professor of Biochemistry. His research areas include mRNA processing in eukaryotes, hormonal regulation of barley α-amylase genes, insect chitinase, and chitin synthase genes and their potential for biocontrol of insects, and plant pathogenesis-related proteins and their defensive roles against plant pathogenic fungi in transgenic plants.

Thorsten Nuernberger obtained an MD in biochemistry in 1987 at Martin-Luther University, Halle-Wittenberg, Germany, and a PhD in plant biochemistry in 1991 at the same institution. During 1991 to

1994 he was a postdoctoral fellow in the Department of Biochemistry, Max-Planck Institute of Plant Breeding, Cologne, Germany. In 1995, he was a postdoctoral fellow at the CNRS Institut des Sciences Vegetales, Gif-sur-Yvette, France. During 1995-2003, he became the Scientific Assistant and Group Leader at the Institute of Plant Biochemistry, Department of Stress and Developmental Biology, Halle, Germany. His research interests include signal perception and signal transduction in the activation of nonhost resistance, and functional characterization of bacterial effector proteins. Since August 2003, he has held the position of Chair of Plant Biochemistry at the Eberhard-Karls-University, Center of Molecular Biology of Plants (ZMBP) in Tuebingen, Germany.

Dierk Scheel studied chemistry and plant biochemistry and obtained a PhD in plant biochemistry from the University of Freiburg, Germany. He continued his work as a postdoctoral fellow in Freiburg and then at the University of California, Berkeley. From 1986 to 1994, he was a vice-director in the Department of Biochemistry, Max-Planck Institute of Plant Breeding in Köln, Germany, and group leader. Since 1994, he has been the Head of the Department of Stress and Developmental Biology at the Institute of Plant Biochemistry, Weinberg, Germany, and a professor of Developmental Biology at Martin Luther University, Halle-Wittenberg, Germany. His research interests are in the mechanisms of plant defense against fungal pathogens.

Hans-Peter Stuible studied biology and biochemistry in Erlangen, Nürnberg, Germany, and obtained a PhD in biochemistry working on the structure and function of bacterial fatty acid synthases with Dr. E. Schweizer. Thereafter, he worked for another year in Erlangen as a postdoctoral fellow. In 1998, he joined the Max-Planck Institute for Plant Breeding Research in Cologne to work on plant defense responses, in particular the structure and function of 4-coumarate:CoA ligase. More recently, he has been involved in studying the role of protein modification (ubiquitination and SUMOylation) and degradation during the hypersensitive response in plants.

N. Westerink studied biology at the University of Utrecht, the Netherlands, in the Laboratory of Molecular Microbiology, and at the Department of Phytopathology and Nematology with the biotechnology company Mogen. She received an MSc in 1997. In 2003, she obtained a PhD working on the *Cladosporium fulvum*-tomato interaction at the Laboratory of Phytopathology at the University of Wageningen. Currently she is a scientist at the Department of Human Retrovirology, Academic Medical Center, University of Amsterdam, the Netherlands.

Foreword

A multitude of fungi and funguslike organisms cause diseases in plants, resulting in significant crop losses and damage to natural ecosystems. Nevertheless, plants can withstand attacks by most fungal pathogens, resulting in one of the central tenets of plant pathology: Most plants are resistant to most plant pathogens. This book addresses in seven highly readable chapters the topic of how plants fend off fungal pathogens. Zamir Punja recruited an impressive team of experts on the subject, resulting in a well-balanced account of the molecular, cellular, and physiological aspects of plant resistance to fungal and oomycete pathogens. This book fulfills the need for a sound overview of the field and is an excellent springboard to the literature. It not only will serve as a beneficial reference to those familiar with the subject but will be equally useful to students and newcomers to the field. Many will appreciate the short historical perspectives included in most chapters, and a good balance between basic and applied aspects of fungal disease resistance research will make the book attractive to a wide range of readers.

This is a timely book. Despite the impressive advances made in understanding fungal-plant interactions, important scientific, social, and commercial challenges remain to be addressed before the humane hope of bringing genetic engineering of fungal resistance to the field can be achieved. The future looks bright though. The pace of basic discovery is accelerating mainly due to the application of novel genome technologies to this field. No one has grasped better the critical importance of a solid and broad knowledge base in the study of fungal pathology than E. C. Large, whose 1940 classic book *The Advent of the Fungi* was reprinted recently with great success (my own copy is a worn-out 1962 Dover edition I snatched years ago from a Berkeley bookstore). Large wrote:

> The best that man could do at any time to defend the health of the hypertrophic agricultural plants that in his cunning he had

sought ought or made, was to apply to the work of rearing them the *whole* of his experience and the whole of his science. (p. 439)

Reading through the diverse chapters of the present book provides a clear view of the modern multifaceted investigations of fungal-plant interactions that expand the tradition defined by Large. It also reinforces a sense of excitement about the years to come and the discoveries to be made.

Sophien Kamoun
The Ohio State University
Ohio Agricultural Research and Development Center,
Wooster

Preface

Throughout history, fungal infection of crop plants has repeatedly resulted in catastrophic harvest failures that have caused major economic and social problems in the affected countries. Intensified research to uncover the genetic and molecular bases of how plants can defend themselves against microbial invasion has led to some remarkable discoveries. These advances have led to the development of crop plants with enhanced resistance to fungal pathogens. In this book, recent developments in the area of fungal disease resistance in plants are highlighted. Details on the cellular responses of resistant plant cells, modes of signal transduction, roles of the hypersensitive response in plants and fungal avirulence factors, induced resistance mechanisms, roles of pathogenesis-related proteins, and genetic engineering approaches to enhance disease resistance are provided. Taken together, these topics cover a breadth not provided anywhere else and will bring the reader up to date on recent developments in this seminal area of crop protection.

Chapter 1

Signal Transduction in Plant Defense Responses to Fungal Infection

Dierk Scheel
Thorsten Nuernberger

INTRODUCTION AND BACKGROUND

Fungal infection of crop plants has repeatedly resulted in catastrophic harvest failures that have caused major economic and social problems in the affected countries. Among the causal agents of infectious crop plant diseases, phytopathogenic fungi and oomycetes play the dominant role (Kamoun, 2001; Knogge, 1998). The potential for serious crop epidemics still persists today, as evidenced by recent outbreaks of diseases caused by rust, mildew, or *Phytophthora* species. In addition to causing food shortages, fungal infection of plants can directly affect the health of humans and livestock through poisoning by toxins. Classical attempts at limiting these losses in agriculture are commonly based on extensive use of fungicides, which in turn causes environmental problems. More recently, genetic engineering of crop plants has allowed expression of new or modified traits, such as enhanced disease resistance (see Chapter 7). It should, however, be stated clearly that genetic engineering for disease resistance is still very much in its infancy. In addition, such approaches are significantly impeded by the tendency of pathogens to develop resistance. Thus intensified research uncovering the molecular basis of both fungal pathogenicity and plant disease resistance is required to better protect crop plants against microbial invasion.

Approximately 10 percent of all fungi and oomycetes have acquired the ability to colonize plants or to cause disease (Knogge, 1998). Although individual plant species are hosts often to only a small number of microbial species, and most pathogens colonize only a few host plant species, virtually all flowering plants are subject to infection by phytopathogenic oomycetes or fungi (Knogge, 1998). Oomycetes, which comprise species of the phytopathogenic genera *Phytophthora, Pythium, Peronospora, Bremia,* and *Plasmopara,* constitute a unique lineage of stramenopile eukaryotes unrelated to true fungi but are closely related to heterokont algae (Govers, 2001; Kamoun, 2001). In spite of the distant phylogenetic relationship between oomycetes and fungi, both classes of microorganisms share a number of common physiological features, including strategies to invade plants (Tyler, 2002). Thus the plant-fungal as well as the plant-oomycete interactions are considered in this chapter.

Phytopathogenic fungi and oomycetes are eukaryotic, carbon-heterotrophic microorganisms which, in order to grow and multiply, must obtain nutrients from their hosts. Fungal infection strategies can be divided into two major categories: (1) biotrophic microorganisms which colonize and grow on living hosts and (2) necrotrophic pathogens that kill host tissue and live on dead material. In many cases, however, attempted fungal penetration and growth on host plants is arrested by a complex, multifaceted plant defense response comprising rigidifications of the plant cell wall at the infection site, the production of antimicrobial plant proteins (see Chapter 5), enzymes and phytoalexins, as well as programmed plant cell death, termed the hypersensitive response (see Chapter 3).

Activation of disease resistance responses in plants occurs at the non-cultivar-specific level (nonhost or species resistance, non-cultivar-specific host resistance) or at the cultivar level (cultivar-specific host resistance) (Cohn et al., 2001; Dangl and Jones, 2001; Heath, 2000; Kamoun, 2001; Nürnberger and Scheel, 2001). Cultivar-specific resistance, which is expressed only by particular plant cultivars against some races of a pathogen species, conforms to the gene-for-gene hypothesis and is genetically determined by complementary pairs of pathogen-encoded avirulence *(Avr)* genes and plant resistance *(R)* genes (Cohn et al., 2001; Dangl and Jones, 2001; Takken and Joosten,

2000; Van der Hoorn et al., 2002) (see Chapter 4). Lack or nonfunctional products of either gene would result in disease.

Most AVR proteins are considered virulence factors required for the colonization of host plants, which, upon recognition by resistant host plant cultivars, act as specific elicitors of plant defense and thereby betray the pathogen to the plant's surveillance system (Cohn et al., 2001; Dangl and Jones, 2001; Van der Hoorn et al., 2002). A simple biochemical interpretation of the gene-for-gene hypothesis implies a receptor/ligand-like interaction between plant *R* gene products and the corresponding pathogen-derived *AVR* gene products. However, isolation and functional characterization of numerous plant *R* genes conferring resistance to a variety of phytopathogenic viruses, bacteria, oomycetes, fungi, nematodes, and insects suggests that the situation is likely to be more complex in many plant-pathogen interactions (Cohn et al., 2001; Dangl and Jones, 2001; van der Biezen and Jones, 1998; Van der Hoorn et al., 2002).

The predominant structural motifs found in R proteins are leucine-rich repeats (LRRs) and coiled coils (CCs), both of which suggest a role in protein-protein interactions. Direct interactions between AVR proteins and R proteins have indeed been demonstrated (reviewed in Cohn et al., 2001; Dangl and Jones, 2001; Nürnberger and Scheel, 2001; Van der Hoorn et al., 2002). However, several studies have provided evidence that LRR-type R proteins constitute components of larger signal-perception complexes and may not necessarily bind directly to their matching AVR proteins (Luderer et al., 2001; Nürnberger and Scheel, 2001). This has led to the "guard hypothesis," which predicts that AVR proteins act as virulence factors that contact their cognate pathogenicity targets in host plants or even nonhost plants, but function only as elicitors of cultivar-specific plant resistance when the complementary R protein is recruited into a functional signal-perception complex (Cohn et al., 2001; Dangl and Jones, 2001; van der Biezen and Jones, 1998; Van der Hoorn et al., 2002). Thus the role of the R protein is to monitor ("guard") AVR-mediated perturbance of cellular functions.

Immunity of an entire plant species (non-host or species resistance) toward most phytopathogenic microorganisms is the predominant form of plant disease resistance (Heath, 2000). Infrequent changes in the host range of phytopathogens are indicative of the stability of species immunity. This resistance type is determined by a defensive network comprising both constitutive barriers and inducible reactions (Heath, 2000; Kamoun, 2001). A large variety of microbe-associated products, referred to as "general elicitors," have been described to trigger defense responses in many plant species in a non-cultivar-specific manner (Boller, 1995; Nürnberger and Brunner, 2002). However, it has remained poorly understood why plants would possess recognition capacities for such "antigenic" signals. Recent publications have unveiled striking similarities in the molecular basis of innate immunity in plants with that known for insects and animals. In 1997, a set of definitions was provided to formalize the description of the components of the mammalian innate immune system (Medzhitov and Janeway, 1997). These authors referred to pathogen surface-derived molecules as pathogen-associated molecular patterns (PAMPs), which bind to pattern recognition receptors and thereby trigger the transcriptional activation of immune response genes and subsequent production of antimicrobial compounds (Boman, 1998).

Innate immune responses have been thoroughly studied in humans, mice, and insects, and it was shown that their molecular basis shows remarkable evolutionary conservation across kingdom borders (Aderem and Ulevitch, 2000; Imler and Hoffmann, 2001; Khush and Lemaitre, 2000; Medzhitov and Janeway, 1997; Underhill and Ozinsky, 2002). PAMPs that have been shown to trigger innate immune responses in various vertebrate and nonvertebrate organisms include the lipopolysaccharide (LPS) fraction of Gram-negative bacteria, peptidoglycans derived from Gram-positive bacteria, eubacterial flagellin, bacterial DNA, as well as fungal cell wall–derived glucans, chitins, mannans, and proteins (Aderem and Ulevitch, 2000; Hemmi et al., 2000; Imler and Hoffmann, 2001; Thoma-Uszynski et al., 2001; Underhill and Ozinsky, 2002). Intriguingly, many of these molecules have long been recognized to be triggers of non-cultivar-specific defense responses in a multitude of plants (Boller, 1995; Dow et al., 2000; Ebel and Scheel, 1997; Heath, 2000; Nürnberger and Brunner, 2002; Tyler, 2002). This suggests that plants, similar to numerous other eukaryotic organisms, have evolved recognition systems for common pathogen-associated surface structures, which initiate intracellular signaling cascades and, ultimately, the activation of protective measures.

Regardless of the type of resistance employed, activation of the plant's surveillance system requires sensitive perception of pathogen-derived molecular patterns or motifs. Because manipulation of signal-transduction pathways is believed to be one of the ways to engineer crop plants with enhanced disease resistance, research in this area has gained enormous momentum in recent years. This chapter reviews recent advances in our understanding of signal perception and transduction events implicated in the establishment of plant disease resistance against phytopathogenic oomycetes and fungi.

SIGNALS AND SIGNAL DELIVERY IN PLANT-FUNGUS/OOMYCETE INTERACTIONS

Research over the past decade has demonstrated that plants have evolved recognition capacities for numerous fungal surface-derived compounds with plant defense-inducing activity, in both host and non-host plants (Boller, 1995; Dow et al., 2000; Nürnberger and Brunner, 2002). These elicitors (termed "general elicitors") comprise (glyco)proteins, peptides, carbohydrates, and lipids, all of which were shown to trigger the activation of plant defense responses. The intrinsic function of such elicitors in the life cycle of the respective fungi or oomycetes remains elusive. Based on their constitutive presence in the cell wall, such elicitors may, however, constitute structural components or enzymes. Examples of enzymes having elicitor activity comprise xylanases from *Trichoderma reesei,* or transglutaminases from *Phytophthora* species (Brunner et al., 2002; Enkerli et al., 1999). Proven independence of elicitor activity from enzymatic activity has indicated that these enzymes may not exert their defense-eliciting effects through the generation of plant-derived "endogenous" elicitors.

Unifying features of PAMPs (inducers of innate immune responses in animals) are their highly conserved structures, their functional importance for and their presence in various microorganisms, and their apparent absence in potential host organisms. Do general elicitors of non-cultivar-specific plant defense responses display such characteristics? Our recent studies (Brunner et al., 2002) revealed that Pep-13 (Nürnberger et al., 1994), a surface-exposed peptide se-

quence present within a novel calcium-dependent cell wall transglut-
aminase, can serve as a recognition determinant for the activation of
plant defense in parsley and potato during interactions with *Phytophthora* species. Pep-13 sequences were found to be highly conserved
among ten *Phytophthora* species analyzed but were virtually absent
in plant sequences. In addition, mutational analysis within the Pep-13
sequence identified amino acid residues indispensable for both trans-
glutaminase activity and the activation of plant defense responses.
This suggests that plants recognize PAMPs with characteristics identical to those triggering innate defense in humans and *Drosophila*.
Activation of plant defense upon recognition of pathogen-associated
structures that are not subject to frequent mutation is likely to provide
a fitness penalty to the pathogen (Kamoun, 2001). Similar to Pep-13,
fungal chitin and oomycete glucans all represent microbe-specific
structures expected to be indispensable for the microbial host (Boller,
1995; Nürnberger and Brunner, 2002).

Plant cells encounter a variety of these signals when interacting
with microorganisms in vivo, and recognition of complex pathogen-
associated molecular patterns is likely to determine the efficiency of
inducible innate defense mechanisms. For example, the cell walls of
many phytopathogenic fungi harbor chitins, *N*-mannosylated glycopeptides, and ergosterol, all of which have been reported to trigger
plant defense responses (Boller, 1995; Ito et al., 1997). Moreover,
phytopathogenic oomycetes of the genera *Phytophthora* and *Pythium*
were shown to possess defense-eliciting heptaglucan structures, elicitins, and other cell wall proteins, such as the necrosis-inducing proteins NPP1, PsojNIP, or PaNie (Fellbrich et al., 2002; Kamoun, 2001;
Mithöfer et al., 2000; Qutob et al., 2002; Umemoto et al., 1997; Veit
et al., 2001). Although not all plant species may recognize and respond to all of these signals, plant cells have recognition systems for
multiple signals derived from individual microbial species. This is
exemplified by tobacco cells, which recognize *Phytophthora*-derived
elicitins and NPP1 (Fellbrich et al., 2002; Kamoun, 2001; Qutob
et al., 2002; Veit et al., 2001), whereas tomato cells were shown to
perceive fungal chitin fragments, glycopeptides, and ergosterol (Basse
et al., 1993; Baureithel et al., 1994; Granado et al., 1995). Taken together, complex pattern recognition by plants is yet another phenomenon reminiscent of the activation of innate defense responses in ani-

mals. For example, innate immune responses in humans are activated by Gram-negative bacteria-derived LPS, flagellin, and unmethylated CpG dinucleotides, which are characteristic of bacterial DNA (Imler and Hoffmann, 2001; Underhill and Ozinsky, 2002). It is yet unclear whether recognition of multiple signals derived from one pathogen may mediate more sensitive perception or if redundant recognition systems may act as independent backup systems (Underhill and Ozinsky, 2002).

In contrast to elicitors of non-plant cultivar-specific defense, AVR proteins are very often synthesized and secreted only upon (attempted) infection of host plants. Thus the genuine function of AVR proteins is likely to promote the colonization of host plants. Resistant host-plant cultivars, which have acquired an *R* gene matching a fungal *Avr* gene, may, however, betray the pathogen to the plant's surveillance system (Cohn et al., 2001; Dangl and Jones, 2001; Takken and Joosten, 2000). A role as a pathogenicity factor has been ascribed to NIP1, a necrosis-inducing peptide from the barley leaf scald-causing fungus, *Rhynchosporium secalis* (Hahn et al., 1993; Rohe et al., 1995). NIP1 was shown to contribute to disease progression on susceptible barley cultivars. NIP1 activity during fungal infection appears to be based on its ability to activate plant plasma membrane H^+-ATPase in all cultivars (Wevelsiep et al., 1993). Fungal races producing NIP1 were shown to trigger race-specific resistance responses on barley cultivars expressing the *Rrs1* resistance gene (Hahn et al., 1993; Rohe et al., 1995). Intriguingly, NIP1 concentrations required to trigger disease-associated lesions in barley are substantially higher than those required for the activation of defense responses in resistant barley cultivars (W. Knogge, personal communication). Apparently, resistant host-plant cultivars may have acquired the ability to recognize a fungal pathogen through tolerable nontoxic concentrations of a fungal toxin. The causal agent of tomato leaf mold, *Cladosporium fulvum*, produces the 28-amino acid polypeptide AVR9, which triggers hypersensitive cell death in tomato plants carrying the *Cf-9* gene (Joosten et al., 1994). In contrast to *R. secalis* NIP1, AVR9 appeared to be dispensable for fungal virulence, suggesting the existence of proteins functionally redundant to AVR9 in *C. fulvum* (Joosten et al., 1994). Potato virus X-based expression of the AVR9-encoding cDNA or infiltration of AVR9 into *Cf-9* tomato cultivars resulted in HR-associated resistance, suggesting that recognition of the AVR protein occurred at the tomato plasma membrane (Joosten et al., 1994; Lauge et al., 2000). In contrast, AvrPita from the rice blast fungus, *Magnaporthe grisea*, was shown to interact in vitro with the matching *R* gene product, Pi-ta, a predicted cytoplasmic rice protein (Jia et al., 2000). This is intriguing since it suggests introduction of a fungal effector protein into the plant cell cytoplasm by a yet unknown secretion/translocation mechanism.

SIGNAL PERCEPTION—A COMPLEX MATTER

Recent findings have significantly improved our understanding of signal perception mechanisms at the plasma membrane of both non-host and host plants. For example, a 75 kDa plasma membrane-bound protein found in various Fabaceae constitutes a binding site for a *Phytophthora sojae*-derived hepta-β-glucan elicitor of phytoalexin production (Mithöfer et al., 2000; Umemoto et al., 1997). Heterologous expression of the encoding soybean gene in tomato conferred high affinity binding of the elicitor. Since this putative receptor protein does not contain recognizable functional domains for signal transmission across the plasma membrane or for intracellular signal generation, it may be recruited into a multicomponent perception complex (Mithöfer et al., 2000). Consistent with this, several proteins closely associated with the heptaglucan binding site were detected in photoaffinity labeling experiments.

Similarly, chemical cross-linking experiments performed with Pep-13 and parsley membranes detected two protein species (100 kDa, 135 kDa) as putative binding proteins. However, as the 100 kDa protein bound Pep-13 in the absence of the 135 kDa protein, their functional interrelationship remains to be elucidated (Nennstiel et al., 1998).

The elicitin receptor represents another example for complex formation implicated in elicitor perception by plants. Elicitins, which constitute a molecular pattern associated with various *Phytophthora* and *Pythium* species (Kamoun, 2001), trigger plant defense in tobacco upon binding to a receptor complex comprising *N*-glycoproteins of 162 and 50 kDa. (Bourque et al., 1999). High-affinity binding sites for elicitins were also reported from *Arabidopsis* and *Acer pseudoplatanus* cells. Elicitins possess the ability to bind sterols, suggesting that the function of these proteins during plant infection is to provide the oomycete with essential lipids (Osman, Mikes, et al., 2001). Recently, it was shown that sterol-elicitin complexes bind more efficiently to the elicitin receptor than elicitins alone, and it was proposed that sterol loading by elicitins may precede binding of the elicitin/sterol complex to the plant receptor (Osman, Vauthrin, et al., 2001). Apparently, the elicitin receptor "guards" against pathogens that use elicitins to manipulate plant sterol homeostasis. Thus, the "guard" hypothesis (Dangl and Jones, 2001; van der Biezen and Jones, 1998; Van der Hoorn et al., 2002) provided to describe AVR/R protein interactions may also explain pathogen recognition processes mediating the activation of non-cultivar-specific plant defense.

Fungal chitin perception is widespread among plant species (Boller, 1995; Day et al., 2001; Ito et al., 1997). A chitinase-related receptor-like kinase, CHRK1, exhibiting autophosphorylation activity but no chitinase activity, was identified in tobacco plasma membranes (Kim, Lee, Yoon, et al., 2000). However, binding of chitin fragments to CHRK1 has yet to be shown. Since CHRK1-encoding transcripts accumulated strongly upon infection with tobacco mosaic virus or *Phytophthora nicotianae*, it is conceivable that CHRK1 might function as a surface receptor for fungus-derived chitin fragments.

The prevalent biochemical interpretation of the gene-for-gene relationship predicts *Avr* gene products to be ligands, which are recognized by *R*-gene-encoded plant receptors. Approximately 20 *R* genes conferring resistance to fungi (*Fusarium oxysporum, Magnaporthe grisea, Puccinia sorghi, Erysiphe graminis, Melampsora lini,* and *Cladosporium fulvum*) and oomycetes *(Peronospora parasitica, Phytophthora infestans)* have been isolated from both monocotyledenous and dicotyledenous plants (Ballvora et al., 2002; reviewed in Takken and Joosten, 2000). The predominant extracellular motifs found in R proteins are leucine-rich repeats which suggest a role in protein-protein interaction (Figure 1.1). Thus R proteins may function in ligand perception. However, only in the case of the cytoplasmic rice protein, Pi-ta, it was shown that *R*-gene-encoded LRR-proteins directly interact with the matching AVR protein, AVR Pi-ta from *Magnaporthe grisea* (Jia et al., 2000). A number of studies have instead revealed that LRR pro-

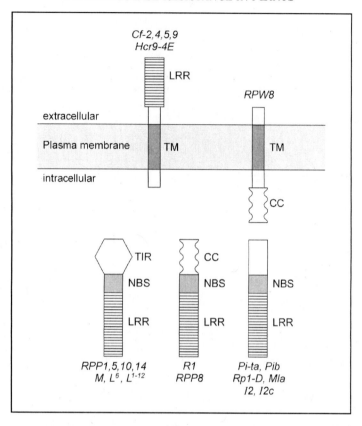

FIGURE 1.1. Structural classes of plant resistance (R) proteins involved in recognition of phytopathogenic fungi and oomycetes (*Source:* Adapted from Ballovara et al., 2002; Takken and Joosten, 2000. *Note:* The corresponding *R* genes are indicated above and below the schematic predicted R protein structures, respectively. LRR, leucine-rich repeat; TM, transmembrane region; CC, coiled-coil structure; TIR, Toll and interleukin-1 receptor domain; NBS, nucleotide-binding site.)

teins apparently constitute components of larger signal perception complexes but do not bind their corresponding AVR proteins directly. These findings support the guard hypothesis (van der Biezen and Jones, 1998; Van der Hoorn et al., 2002), which predicts that AVR proteins bind to their cognate pathogenicity targets in host plants or even nonhost plants, and that the *R* gene product merely guards perturbations of cellular homoeostasis. The tomato *Cf-9* resistance gene encodes a membrane-anchored glycoprotein with an extracellular LRR and a small cytoplasmic domain without apparent function in downstream signaling (Jones et al., 1994; Piedras et al., 2000). Recent studies on the topology of the Cf-9 protein revealed its presence in the plasma membrane (Jones et al., 1994; Piedras et al., 2000; Rivas et al., 2002) as well as in the endoplasmic reticulum (Benghezal et al., 2000). Moreover, since both susceptible and resistant cultivars of tomato as well as other solanaceous plants harbor a high-affinity binding site for AVR9, the Cf-9 protein is unlikely to be the AVR9 receptor (Kooman-Gersmann et al., 1996). Consistently, comprehensive biochemical analyses failed to demonstrate a physical interaction of the two proteins (Luderer et al., 2001).

Taken together, plant perception systems for pathogen-associated molecular motifs (general elicitors, AVR proteins) often appear to be complexes rather than single receptor proteins that mediate ligand binding, signal transmission across the plant plasma membrane, and subsequent initiation of an intracellular signaling cascade. This is reminiscent of PAMP recognition by animal receptor complexes which mediate the activation of innate immune responses. In these systems, ligand-binding proteins are recruited into complexes with transmembrane Toll-like receptors only in the ligand-bound form. Plant transmembrane R proteins and LRR-containing Toll-like receptors in animals seem to be structurally conserved, suggesting that LRR proteins constitute the predominant structural basis for pathogen perception in eukaryotes. Moreover, parallels between plant R proteins and animal receptors for pathogen-associated molecular patterns extend to intracellular signaling domains as well (Takken and Joosten, 2000; Underhill and Ozinsky, 2002). For example, plant R proteins (flax L[6], *Arabidopsis* RPP5) (Takken and Joosten, 2000) conferring resistance to *Melampsora lini* or *Peronospora parasitica,* respectively, were shown to carry a cytoplasmic TIR (Toll/interleukin-1-receptor) domain similar to that found in mammalian Toll-like receptors.

CELLULAR SIGNALING—A MATTER OF NETWORKS

The signal transduction networks linking receptor-mediated perception of pathogens and defense reactions employ second messengers that are conserved among most eukaryotes. In particular, second messengers of mammalian innate immunity, such as Ca^{2+}, reactive oxygen species (ROS), nitric oxide (NO), and mitogen-activated protein kinase (MAPK) cascades, are also involved in defense signaling in plants (Nürnberger and Scheel, 2001).

Alterations of ion fluxes across the plasma membrane (Ca^{2+} and H^+ influx, K^+ and Cl^- efflux) were found to be common early events in defense signaling upon fungal and oomycete attack or perception of respective elicitors (Blatt et al., 1999; Jabs et al., 1997; Klüsener and Weiler, 1999; Nürnberger and Scheel, 2001; Scheel, 1998; Zimmermann et al., 1997). Receptor-mediated regulation of specific ion channels appears to be responsible for this transiently altered ion permeability of the plasma membrane (Blatt et al., 1999; Zimmermann et al., 1997), which is necessary and sufficient for the activation of defense reactions (Jabs et al., 1997). Treatment of tobacco guard cells expressing the *Cf-9* resistance gene from tomato with the corresponding AVR9 avirulence protein from *Cladosporium fulvum* activated an outward-rectifying K^+ channel and simultaneously inactivated an inward-rectifying K^+ channel (Blatt et al., 1999). This protein kinase-dependent differential regulation of two K^+ channels can be explained by the frequently observed net K^+ efflux from elicitor-treated plant cells (Ebel and Scheel, 1997; Jabs et al., 1997). Plasma membrane-located Ca^{2+} channels were found to be responsive to the general oomycete elicitor, Pep-13 (Zimmermann et al., 1997), and to race-specific elicitors from *Cladosporium fulvum* (Gelli et al., 1997). Receptor-mediated increases in open probability of these channels caused transient elevation of cytosolic Ca^{2+} levels (Blume et al., 2000; Lecourieux et al., 2002; Mithöfer et al., 1999). Since protein kinase inhibitors block these increases of cytosolic Ca^{2+} concentrations, receptor-mediated regulation of the corresponding Ca^{2+} channels involves protein phosphorylation (Blume et al., 2000; Lecourieux et al., 2002), as shown for the elicitor-responsive K^+ channels (Blatt et al., 1999). Although pharmacological analyses demonstrated the requirement of Ca^{2+} influx from the extracellular space, the participation of internal stores in elevating cytosolic Ca^{2+} levels cannot be ruled out (Blume et al., 2000; Lecourieux et al., 2002; Mithöfer et al., 1999; Xu and Heath, 1998). Amplitude and duration of these defense-related Ca^{2+} transients vary, but prolonged, modest increases of cytosolic Ca^{2+} levels rather than spikes of large intensity or oscillations appear to be essential for elicitation of defense responses (Blume et al., 2000; Lecourieux et al., 2002; Xu and Heath, 1998).

Two structurally related elicitor-responsive calcium-dependent protein kinases (CDPKs) were identified in tobacco (Romeis et al., 2000, 2001), which represent an early downstream target of cytosolic Ca^{2+}. Treatment of transgenic tobacco cells expressing the tomato *Cf-9* resistance gene with the corresponding AVR9 elicitor from *Cladosporium fulvum* resulted in protein phosphorylation, enzyme activation, and increased transcript accumulation of both CDPKs (Romeis et al., 2000, 2001). Since virus-induced gene silencing of the corresponding genes compromised AVR9-stimulated programmed cell death, these CDPKs are involved in activating the plant's defense response (Romeis et al., 2001). The targets of plant CDPKs remain to be identified but may be related to animal protein kinase C, which is involved in activation of the NADPH oxidase of mammalian neutrophils (Jones, 1994). Although plant homologues of the corresponding protein kinase C substrate, the cytosolic NADPH oxidase subunit, p47, have not yet been found, it is interesting to note that expression of an *Arabidopsis* CDPK in tomato protoplasts stimulated NADPH oxidase activity and ROS production (Xing et al., 2001). Together with the finding that elicitor activation of the two tobacco CDPKs is independent of the oxidative burst (Romeis et al., 2000), these data may suggest the involvement of CDPKs in stimulating ROS production.

The involvement of the universal Ca^{2+} signal messenger, calmodulin, in plant defense signaling has only recently been documented. Treatment of suspension-cultured soybean cells with elicitor preparations from *Fusarium solani* and *Phytophthora parasitica* var. *nicotianae* resulted in rapid Ca^{2+}-dependent accumulation of the two calmodulin isoforms, SCaM-4 and SCaM-5, and the corresponding transcripts preceding the salicylate-independent initiation of defense reactions (Heo et al., 1999). Both isoforms were barely detectable in healthy plants, whereas other calmodulins were found to be constitutively expressed. Tobacco plants constitutively expressing either SCaM-4 or SCaM-5 displayed constitutive *PR* (pathogenesis-related) gene expression and enhanced resistance against *Phytophthora parasitica* var. *nicotianae* in a salicylate-independent manner.

Protein-protein interaction-based screening of a cDNA expression library from suspension-cultured rice cells treated with an elicitor preparation from the fungal rice pathogen, *Magnaporthe grisea,* with the soybean calmodulin isoform, SCaM-1, resulted in the isolation of *OsMlo* (Kim, Lee, Park, et al., 2000; Kim, Lee, et al., 2002). *OsMlo* encodes a plasma membrane-located protein with strong homology to barley Mlo (Kim, Lee, et al., 2002). Recessive, presumably loss-of-regulatory-function mutations in barley *Mlo* confer stable resistance against the powdery mildew fungus, *Blumeria graminis* f. sp. *hordei* (Jorgensen, 1992). Expression of *OsMlo* was strongly induced by cocultivation of suspension-cultured rice cells with spores of *Magnaporthe grisea* and in response to treatments with oligochitin elcitor, salicylate, ethephon, jasmonate, and H_2O_2 (Kim, Lee, et al., 2002). A 20-amino acid calmodulin-binding domain was identified within the cytoplasmic C-terminal tail of *OsMlo,* which has seven transmembrane domains. Interestingly, a functional calmodulin-binding domain was also detected in barley *Mlo* and found to be important for *Mlo* function, i.e., suppression of defense against powdery mildew (Kim, Panstruga, et al., 2002). In contrast to animal and fungal seven-transmembrane orthologs, *Mlo* disease signaling does not require heterotrimeric G proteins but involves Ca^{2+}-dependent binding of calmodulin to its C-terminal cytoplasmic tail. Specific calmodulin isoforms, therefore, appear to constitute essential elements of signaling networks regulating initiation as well as suppression of plant defense responses.

ROS production is an important early component of innate immunity in animals and plants (Scheel, 2002a). Extracellular generation of ROS during the oxidative burst of plants depends on transient increases of cytosolic Ca^{2+} levels (Blume et al., 2000) and appears to be mechanistically similar to the respiratory burst of human phagocytes, which is catalyzed by an NADPH oxidase protein complex (Babior et al., 1997; Scheel, 2002a). Plants harbor a family of genes with significant homology to the human gene encoding the catalytic subunit, gp91, of the NADPH oxidase complex (Groom et al., 1996; Keller et al., 1998; Torres et al., 1998). In comparison with the human ortholog, the plant proteins carry an N-terminal extension comprising an EF-hand motif indicative of Ca^{2+} regulation (*Arabidopsis* Genome Initiative, 2000; Keller et al., 1998). Microsomal fractions from yeast expressing a parsley gp91 homologue indeed catalyzed the formation of $O_2^{\cdot-}$ in a Ca^{2+}- and NADPH-dependent manner, suggesting that this protein by itself catalyzes ROS production during pathogen defense (Zinecker, 2001). In tobacco, direct activation of an NADPH oxidase by increased Ca^{2+} levels has been described (Sagi andFluhr, 2001). Antisense silencing of the tobacco plasma membrane-localized NADPH oxidase, NtRbohD, resulted in the complete loss of the oxidative burst response of the transgenic plants upon treatment with the *Phytophthora cryptogea* elicitor, cryptogein (Simon-Plas et al., 2002). Individual and double knock-out mutants of the two major leaf NADPH oxidases of *Arabidopsis thaliana* were strongly impaired in ROS production in incompatible interactions with the oomycete pathogen *Peronospora parasitica* (Torres et al., 2002). Although the knock-out lines displayed an as yet unexplained enhanced cell death response after infection with *P. parasitica,* these data unequivocally demonstrate a causal link between ROS production and plant gp91 orthologs.

Activation of the human respiratory burst oxidase complex involves the small GTP-binding protein Rac2 (Babior et al., 1997). Since orthologs of this gene family are present in plants (Kawasaki et al., 1999), these might be involved in regulating ROS production during plant defense. Transgenic rice plants and cultured cells expressing a constitutively active derivative of OsRac1, one out of the three Rac2 orthologs of rice, produced elevated ROS and phytoalexin levels, developed symptoms of programmed cell death, and showed increased resistance against virulent *Magnaporthe grisea* (Kawasaki et al., 1999; Ono et al., 2001). Expression of the dominant-negative OsRac1 derivative suppressed elicitor-stimulated ROS production and pathogen-induced cell death in transgenic rice (Ono et al., 2001). Additional studies with the rice dwarf mutant, *d1,* that lacks the single Gα gene, demonstrated that this subunit of heterotrimeric G proteins acts upstream of OsRac1 in rice disease resistance (Suharsono et al., 2002). Not only were defense responses of the *d1* mutant to *Magnaporthe grisea* delayed and suppressed, but expression of the constitutively active OsRac1 derivative in *d1* plants restored rice blast resistance and elicitor responsiveness.

In concert with ROS, NO plays an important role in innate immunity of animals and plants (Wendehenne et al., 2001). NO was found to be produced by tobacco, soybean, and *Arabidopsis* upon infection with avirulent bacteria or viruses, but also upon treatment with elicitors, such as cryptogein from *Phytophthora cryptogea* (Clarke et al., 2000; Delledonne et al., 1998; Durner et al., 1998; Foissner et al., 2000). The synthesis of NO in plants is still a matter of debate (Wendehenne et al., 2001), although NO synthase inhibitors blocked infection- and elicitor-stimulated NO production, cell death, and defense-gene activation (Delledonne et al., 1998; Durner et al., 1998). Downstream signaling of NO appears to involve cyclic GMP and cyclic ADP ribose, which then stimulate another cytosolic Ca^{2+} transient (Durner et al., 1998). This late Ca^{2+} response may be caused by activation of the *Arabidopsis* cyclic nucleotide-gated ion channel, AtCNGC2, which harbors a functional calmodulin-binding domain within its C-terminal domain, mediates fluxes of Ca^{2+}, K^+, and other cations, and appears to be involved in initiating programmed cell death (Clough et al., 2000; Köhler et al., 1999, 2001). *AtCNGC2* is affected in the *dnd1* mutant, which fails to develop programmed cell death upon infection with avirulent pathogens, but still displays resistance (Clough et al., 2000).

Aconitase may represent a possible direct target of NO in plants that is involved in the cell death response (Navarre et al., 2000; Wendehenne et al., 2001). In mammals, NO binds to the cytosolic aconitase and converts it to the so-called iron regulatory protein (Wendehenne et al., 2001). This RNA-binding protein modulates the translation of specific mRNAs and thereby causes increased cytosolic free iron levels. As in animals, NO also modulates aconitase activity in plants, and residues involved in regulatory mRNA binding of the human enzyme were conserved in cytosolic tobacco aconitase (Navarre et al., 2000). Although mRNA regulatory activity of this aconitase remains to be demonstrated, the similarities between the mammalian and plant innate immunity systems may suggest an analogous function (Wendehenne et al., 2001). Thus in concert with the oxidative burst, increased free cytosolic iron levels could promote cell death via the Fenton reaction, which generates hydroxyl radicals. Interestingly, NO also induced transcipt accumulation of *AOX1a*, encoding an alternative oxidase, and stimulated respiration through the alternative pathway in *Arabidopsis* (Huang et al., 2002). Inhibition of this respiration by AOX inhibitors increased NO sensitivity and death of the *Arabidopsis* cells, suggesting that NO stimulates protective measures as is also found for ROS (Huang et al., 2002; Scheel, 2002a). Antisense suppression of mitochondrial AOX in tobacco rendered these cells hypersensitive to initiation of programmed cell death by salicylic acid and H_2O_2, again suggesting that AOX is involved in protecting mitochondria from oxidative stress as it may originate from the oxidative burst (Robson and Vanleberghe, 2002).

MAPK cascades are believed to represent a central point of crosstalk in stress signaling in plants (Ichimura et al., 2002; Nürnberger and Scheel, 2001; Zhang and Klessig, 2001). MAPKs are the most downstream components of these cascades and consist of at least three protein kinases that are sequentially activated by phosphorylation. Based on the complete genome sequence of *Arabidopsis thaliana*, this plant harbors at least 20 MAPKs that are activated via dual phosphorylation of a typical threonine/X/tyrosine motif by a maximum of ten MAPK kinases (MAPKK) (Ichimura et al., 2002; *Arabidopsis* Genome Initiative, 2000). The number and structural variation of MAPKK kinases (MAPKKK) that activate MAPKKs is even greater than that of MAPKs (Ichimura et al., 2002). Fungal and oomycete elicitors activate primarily MAPKs of the *Arabidopsis* MPK3 and MPK6 type (Cardinale et al., 2000; Ichimura et al., 2002; Kroj et al., 2003; Romeis et al., 1999; Zhang and Klessig, 1998, 2001), although activation of AtMPK3 itself by elicitor treatment or infection has not been detected (Nühse et al., 2000). However, *AtMPK3* transcript accumulated rapidly upon treatment of *Arabidopsis* seedlings with chitin or chitin oligomers (Zhang et al., 2002). MAPK activation is located downstream of the elicitor-stimulated Ca^{2+} influx and appears not to be necessary for the oxidative burst (Ligterink et al., 1997; Romeis et al., 1999; Yang et al., 2001). In elicitor-treated parsley cells, at least one of the activated MAPKs translocates to the nucleus (Ligterink et al., 1997), where it is involved in oxidative burst-independent activation of *PR* genes (Kroj et al., 2003) that had previously been shown to be regulated by WRKY transcription factors (Eulgem et al., 1999). Interestingly, a complete *Arabidopsis* MAPK cascade has recently been identified that is involved in innate immunity via a WRKY transcription factor and mediates increased resistance to fungal and bacterial pathogens (Asai et al., 2002). The existence of a causal link between MAPK activation and defense-gene activation, as well as programmed cell death, was furthermore suggested by gain-of-function experiments transiently overexpressing MAPKs themselves or constitutively active derivatives of the corresponding MAPKKs in tobacco and *Arabidopsis* leaves, which caused a programmed cell death-like phenotype and activation of defense-related genes (Ren et al., 2002; Yang et al., 2001; Zhang and Liu, 2001).

Other MAPK cascades were found to be negative regulators of plant defense. *Arabidopsis* mutants carrying a transposon insertion in the *AtMPK4* gene had an extreme dwarf phenotype and exhibited elevated salicylate levels, constitutive *PR* gene expression, and increased resistance against virulent pathogens, including *Peronospora parasitica* (Petersen et al., 2000). This resistance was independent of NPR1 but required salicylate. The *edr1* mutant of *Arabidopsis* also showed increased resistance against bacterial and fungal pathogens, such as powdery mildew, but otherwise displayed a normal phenotype (Frye et al., 2001). *Edr1* encodes a putative Raf-like MAPKKK similar to the negative regulator of the ethylene response, CTR1 (Johnson and Ecker, 1998). In contrast to *mpk4, edr1*-mediated resistance depended on salicylate and NPR1 and did not show enhanced salicylate levels and constitutive *PR* gene expression (Frye et al., 2001). It is therefore unlikely that the corresponding MAPKKK is part of the AtMPK4 cascade.

Posttranslational protein modification by covalent attachment of ubiquitin or ubiquitin-like proteins plays a critical role in the regulation of many cellular processes, among them innate immunity (Hershko and Ciechanover, 1998; Yeh et al., 2000). The central importance of these regulatory tools in plant defense signaling has only recently been recognized in *R*-gene-mediated defense against fungi and oomycetes, which involve *EDS1, PAD4, NDR1,* and *RAR1* (Austin et al., 2002; Azevedo et al., 2002; Tör et al., 2002). Eds1 and Pad4 are lipaselike proteins necessary for disease resistance mediated by R proteins with TIR domains (Glazebrook, 2001). Ndr1 is involved in defense signaling initiated by CC-domain R proteins and is predicted to be a glycosylphosphatidylinositol-anchored membrane protein (Glazebrook, 2001). *RAR1* is required for powdery mildew resistance conveyed by different R proteins, including Mla6 and Mla12 (Shirasu et al., 1999). The encoded protein harbors the two zinc-binding domains, CHORD-I and CHORD-II, and interacts via CHORD-II with plant orthologs of the yeast SGT1 protein, which plays a role in ubiquitin-mediated protein degradation (Azevedo et al., 2002; Kita-gawa et al., 1999). SGT1 and the RAR1-SGT1 complex co-immuno-precipitated with CSN4 and CSN5, two subunits of COP9, which directly interact with SCF-type E3 ubiquitin ligases (Azevedo et al., 2002; Lyapina et al., 2001; Schwechheimer et al., 2001). Silencing of *SGT1* in barley impaired *Mla6*- but not *Mla1*-mediated resistance against *Blumeria graminis* f. sp. *hordei* (Azevedo et al., 2002). Whereas barley appears to harbor only a single *SGT1* gene, there are two copies in *Arabidopsis, SGT1a* and *SGT1b* (Azevedo et al., 2002). Only *SGT1b* was found to be involved in *R*-gene-conferred resistance against different isolates of *Peronospora parasitica* (Austin et al., 2002; Tör et al., 2002). In some of these interactions, the *Arabidopsis RAR1* ortholog and in others, *NDR1,* were also necessary for resistance. Interestingly, silencing of *SGT1* in *Nicotiana benthamiana* also compromised the stimulation of programmed cell death by the elicitin INF1 from *Phytophthora infestans* (Peart et al., 2002). These findings strongly suggest that ubiquitin-mediated protein degradation is an important early component of plant defense signal transduction, where it might function in the removal of regulators.

The signaling elements described regulate either salicylate- or jasmonate/ethylene-mediated responses (Glazebrook, 2001), which themselves are interlinked by elements that affect both types of signaling (Kunkel and Brooks, 2002). The MAPK MPK4, for example, is a negative regulator of salicylate-mediated signaling but promotes jasmonate-conveyed responses (Petersen et al., 2000). Such signaling elements may be necessary for cross talk between different signal-transduction networks, which allows the fine-tuning of the plant's response to the multitude of environmental stimuli (Kunkel and Brooks, 2002).

Several genes encoding proteins involved in pathogen recognition and defense signaling have been employed in biotechnological approaches to generate plants with improved disease resistance (Scheel, 2002b) (see Chapter 7). Although commercial products have not yet been released, overexpression studies with model systems clearly demonstrate that diverse signaling elements can be used to engineer increased resistance without significantly affecting the normal phenotype. Transgenic *Arabidopsis* plants with elevated methyl jasmonate levels and tobacco plants with increased salicylate, for example, displayed higher levels of resistance to fungal pathogens than the corresponding wild-type plants (Seo et al., 2001; Verberne et al., 2000). Interestingly, overexpression of the *Pto* gene, normally conferring resistance against *Pseudomonas syringae* pv. *tomato,* also yielded increased resistance against *Cladosporium fulvum* (Tang et al., 1999). Expression of corresponding pairs of *R* and *Avr* genes, one of which was under control of a strictly pathogen-responsive promoter, also led to plants with improved broad resistance against fungal pathogens (Hennin et al., 2001; Melchers and Stuiver, 2000). Pathogen-inducible production of the elicitin cryptogein in tobacco plants, which harbor the corresponding receptor, rendered these plants resistant toward a broad spectrum of fungal and oomycete pathogens (Keller et al., 1999), whereas constitutive expression of specific calmodulin isoforms in tobacco resulted in enhanced resistance against *Phytophthora parasitica* var. *nicotianae* (Heo et al., 1999). Several transcription factors that regulate the expression of defense-related genes have successfully been employed to improve disease resistance (Scheel, 2002b). Furthermore, expression of constitutively active derivatives of signaling elements, such as G proteins or different kinases, might also offer tools for the production of disease-resistant plants (Kawasaki et al., 1999; Ono et al., 2001; Ren et al., 2002; Romeis et al., 2001; Yang et al., 2001; Zhang and Liu, 2001). Thus a better understanding of defense signaling will provide new tools for engineering fungal disease resistance in plants.

CONCLUSIONS

Pathogen recognition and subsequent signal transduction in plants is mediated by components that are structurally and functionally similar to the innate immune signaling system of mammals and insects (Aderem and Ulevitch, 2000; Khush and Lemaitre, 2000; Nürnberger and Scheel, 2001). Plant receptor complexes with conserved functional domains perceive microbial elicitors of identical or similar structure as PAMPs that are recognized by Toll-like receptors and mediate activation of innate immune responses in animals (Nürnberger and Brunner, 2002). The subsequently initiated signaling networks employ some of the same second messengers that are involved in the activation of innate immune responses in animals as well. Moreover, even the complex defense responses include a few functionally similar elements, such as antimicrobial defensins (Thomma et al., 2002). Microarray analyses of chitin elicitation in *Arabidopsis thaliana* revealed that the expression of at least 61 out of 2,375 genes tested was altered threefold or more relative to control plants (Ramonell et al., 2002). Comparison of this set of genes to those activated by infection with *Alternaria brassicicola* in *Arabidopsis* plants resistant to this fungal pathogen revealed that roughly 50 percent of the genes responded in the same manner to both treatments (Ramonell et al., 2002; Schenk et al., 2000). Thus a single elicitor can activate a complex, stereotyped gene set that shows significant overlap with that activated upon recognition of complete microbial pathogens, a phenomenon which has also been reported from the activation of innate immune responses in human cells (Boldrick et al., 2002). Taken together, these obvious parallels in the molecular organization of innate immunity in both animals and plants support the hypothesis that they might have a common evolutionary origin.

REFERENCES

Aderem, A. and Ulevitch, R. (2000). Toll-like receptors in the induction of the innate immune response. *Nature* 406: 782-787.

The *Arabidopsis* Genome Initiative (2000). Analysis of the genome sequence of the flowering plant *Arabidopsis thaliana*. *Nature (London)* 408: 796-815.

Asai, T., Tena, G., Plotnikova, J., Willmann, M.R., Chiu, W.-L., Gomez-Gomez, L., Boller, T., Ausubel, F.M., and Sheen, J. (2002). MAP kinase signalling cascade in *Arabidopsis* innate immunity. *Nature* 415: 977-983.

Austin, M.J., Muskett, P., Kahn, K., Feys, B.J., Jones, J.D.G., and Parker, J.E. (2002). Regulatory role of *SGT1* in early *R* gene-mediated plant defenses. *Science* 295: 2077-2080.

Azevedo, C., Sadanandom, A., Kitagawa, K., Freialdenhoven, A., Shirasu, K., and Schulze-Lefert, P. (2002). The RAR1 interactor SGT1, an essential component of *R* gene-triggered disease resistance. *Science* 295: 2073-2076.

Babior, B.M., Benna, J.E., Chanock, S.J., and Smith, R.M. (1997). The NADPH oxidase of leukocytes: The respiratory burst oxydase. In Scandalios, J.G. (Ed.), *Oxidative Stress and the Molecular Biology of Antioxidant Defenses,* Volume 34 (pp. 737-783). Cold Spring Harbor, NY: Cold Spring Harbor Laboratory Press.

Ballvora, A., Ercolano, M.R., Weiss, J., Meksem, K., Bormann, C.A., Oberhagemann, P., Salamini, F., and Gebhardt, C. (2002). The *R1* gene for potato resistance to late blight *(Phytophthora infestans)* belongs to the leucine zipper/NBS/LRR class of plant resistance genes. *The Plant Journal* 30: 361-371.

Basse, C.W., Fath, A., and Boller, T. (1993). High affinity binding of a glycopeptide elicitor to tomato cells and microsomal membranes and displacement by specific glycan suppressors. *Journal of Biological Chemistry* 268: 14724-14731.

Baureithel, K., Felix, G., and Boller, T. (1994). Specific, high affinity binding of chitin fragments to tomato cells and membranes. *Journal of Biological Chemistry* 269: 17931-17938.

Benghezal, M., Wasteneys, G.O., and Jones, D.A. (2000). The C-terminal dilysine motif confers endoplasmic reticulum localisation to type I membrane proteins in plants. *The Plant Cell* 12: 1179-1201.

Blatt, M.R., Grabov, A., Brearley, J., Hammond-Kosack, K., and Jones, J.D.G. (1999). K$^+$ channels of *Cf-9* transgenic tobacco cells as targets for *Cladosporium fulvum Avr9* elicitor-dependent signal transduction. *The Plant Journal* 19: 453-462.

Blume, B., Nürnberger, T., Nass, N., and Scheel, D. (2000). Receptor-mediated increase in cytoplasmic free calcium required for activation of pathogen defense in parsley. *The Plant Cell* 12: 1425-1440.

Boldrick, J.C., Alizadeh, A.A., Diehn, M., Dudoit, S., Liu, C.L., Belcher, C.E., Botstein, D., Staudt, L.M., Brown, P.O., and Relman, D.A. (2002). Stereotyped and specific gene expression programs in human innate immune responses to bacteria. *Proceedings of the National Academy of Sciences, USA* 99: 972-977.

Boller, T. (1995). Chemoperception of microbial signals in plant cells. *Annual Review of Plant Physiology and Plant Molecular Biology* 46: 189-214.

Boman, H.G. (1998). Gene-encoded peptide antibiotics and the concept of innate immunity: An update review. *Scandinavian Journal of Immunology* 48: 15-25.

Bourque, S., Binet, M.-N., Ponchet, M., Pugin, A., and Lebrun-Garcia, A. (1999). Characterization of the cryptogein binding sites on plant plasma membranes. *Journal of Biological Chemistry* 27: 34699-34705.

Brunner, F., Rosahl, S., Lee, J., Rudd, J.J., Geiler, S., Kauppinen, S., Rasmussen, G., Scheel, D., and Nürnberger, T. (2002). A PAMP for the activation of plant innate immune responses to pathogenic *Phytophthora*. *The EMBO Journal* 21: 6681-6688.

Cardinale, F., Jonak, C., Ligterink, W., Niehaus, K., Boller, T., and Hirt, H. (2000). Differential activation of four specific MAPK pathways by distinct elicitors. *Journal of Biological Chemistry* 275: 36734-36740.

Clarke, A., Desikan, R., Hurst, R.D., Hancock, J.T., and Neill, S.J. (2000). NO way back: Nitric oxide and programmed cell death in *Arabidopsis thaliana* suspension cultures. *The Plant Journal* 24: 667-677.

Clough, S.J., Fengler, K.A., Yu, I.C., Lippok, B., Smith, R.K., and Bent, A.F. (2000). The *Arabidopsis dnd1* "defense, no death" gene encodes a mutated cyclic nucleotide-gated ion channel. *Proceedings of the National Academy of Sciences, USA* 97: 9323-9328.

Cohn, J., Sessa, G., and Martin, G.B. (2001). Innate immunity in plants. *Current Opinion in Immunology* 13: 55-62.

Dangl, J.L. and Jones, J.D.G. (2001). Plant pathogens and integrated defence responses to infection. *Nature* 411: 826-833.

Day, R.B., Okada, M., Ito, Y., Tsukada, K., Zaghouani, H., Shibuya, N., and Stacey, G. (2001). Binding site for chitin oligosaccharides in the soybean plasma membrane. *Plant Physiology* 126: 1162-1173.

Delledonne, M., Xia, Y., Dixon, R., and Lamb, C. (1998). Nitric oxide functions as a signal in plant disease resistance. *Nature* 394: 585-588.

Dow, M., Newman, M.-A., and von Roepenack, E. (2000). The induction and modulation of plant defense responses by bacterial lipopolysaccharides. *Annual Review of Phytopathology* 38: 241-261.

Durner, J., Wendehenne, D., and Klessig, D.F. (1998). Defense gene induction in tobacco by nitric oxide, cyclic GMP and cyclic ADP ribose. *Proceedings of the National Academy of Sciences, USA* 95: 10328-10333.

Ebel, J. and Scheel, D. (1997). Signals in host-parasite interactions. In Carroll, G.C. and Tudzynski, P. (Eds.), *The Mycota: Plant Relationships*, Part A, Volume V (pp. 85-105). Berlin, Heidelberg: Springer-Verlag.

Enkerli, J., Felix, G., and Boller, T. (1999). The enzymatic activity of fungal xylanase is not necessary for its elicitor activity. *Plant Physiology* 121: 391-397.

Eulgem, T., Rushton, P.J., Schmelzer, E., Hahlbrock, K., and Somssich, I.E. (1999). Early nuclear events in plant defence signalling: Rapid activation by WRKY transcription factors. *The EMBO Journal* 18: 4689-4699.

Fellbrich, G., Romanski, A., Varet, A., Blume, B., Brunner, F., Engelhardt, S., Felix, G., Kemmerling, B., Krzymowska, M., and Nürnberger, T. (2002). NPP1, a *Phytophthora*-associated trigger of plant defense in parsley and *Arabidopsis*. *The Plant Journal* 32: 375-390.

Foissner, I., Wendehenne, D., Langebartels, C., and Durner, J. (2000). In vivo imaging of an elicitor-induced nitric oxide burst in tobacco. *The Plant Journal* 23: 817-824.

Frye, C.A., Tang, D., and Innes, R.W. (2001). Negative regulation of defense responses in plants by a conserved MAPKK kinase. *Proceedings of the National Academy of Sciences, USA* 98: 373-378.

Gelli, A., Higgins, V.J., and Blumwald, E. (1997). Activation of plant plasma membrane Ca^{2+}-permeable channels by race-specific fungal elicitors. *Plant Physiology* 113: 269-279.

Glazebrook, J. (2001). Genes controlling expression of defense responses in *Arabidopsis*—2001 status. *Current Opinion in Plant Biology* 4: 301-308.

Govers, F. (2001). Misclassification of pest as "fungus" puts vital research on wrong track. *Nature* 411: 633.

Granado, J., Felix, G., and Boller, T. (1995). Perception of fungal sterols in plants. *Plant Physiology* 107: 485-490.

Groom, Q.J., Torres, M.A., Fordham-Skelton, A.P., Hammond-Kosack, K.E., Robinson, N.J., and Jones, J.D.G. (1996). *rbohA*, a rice homologue of the mammalian *gp91phox* respiratory burst oxidase. *The Plant Journal* 10: 515-522.

Hahn, M., Jungling, S., and Knogge, W. (1993). Cultivar-specific elicitation of barley defense reactions by the phytotoxic peptide NIP1 from *Rhynchosporium secalis*. *Molecular Plant-Microbe Interactions* 6: 745-754.

Heath, M.C. (2000). Nonhost resistance and nonspecific plant defenses. *Current Opinion in Plant Biology* 3: 315-319.

Hemmi, H., Takeuchi, O., Kawai, T., Kaisho, T., Sato, S., Sanjo, H., Matsumoto, M., Hoshino, K., Wagner, H., Takeda, K., and Akira, S. (2000). A Toll-like receptor recognizes bacterial DNA. *Nature* 408: 740-745.

Hennin, C., Höfte, M., and Diederichsen, E. (2001). Functional expression of *Cf9* and *Avr9* genes in *Brassica napus* induces enhanced resistance to *Leptosphaeria maculans*. *Molecular Plant-Microbe Interactions* 14: 1075-1085.

Heo, W.D., Lee, S.H., Kim, M.C., Kim, J.C., Chung, W.S., Chun, H.J., Lee, K.J., Park, C.Y., Park, H.C., Choi, J.Y., and Cho, M.J. (1999). Involvement of specific calmodulin isoforms in salicylic acid-independent activation of plant disease resistance responses. *Proceedings of the National Academy of Sciences, USA* 96: 766-771.

Hershko, A. and Ciechanover, A. (1998). The ubiquitin system. *Annual Review of Biochemistry* 67: 425-479.

Huang, X., von Rad, U., and Durner, J. (2002). Nitric oxide induces transcriptional activation of the nitric oxide tolerant alternative oxidase in *Arabidopsis* suspension cells. *Planta* 215: 914-923.

Ichimura, K., Tena, G., Henry, Y., Zhang, S., Hirt, H., Ellis, B.E., Morris, P.C., Wilson, C., Champion, A., Innes, R.W., et al. (2002). Mitogen-activated protein kinase cascades in plants: A new nomenclature. *Trends in Plant Science* 7: 301-308.

Imler, J.-L. and Hoffmann, J.A. (2001). Toll receptors in innate immunity. *Trends in Cell Biology* 11: 304-311.

Ito, Y., Kaku, H., and Shibuya, N. (1997). Identification of a high-affinity binding protein for *N*-acetylchitoologosaccharide elicitor in the plasma membrane of suspension-cultured rice cells by affinity labeling. *The Plant Journal* 12: 347-356.

Jabs, T., Tschöpe, M., Colling, C., Hahlbrock, K., and Scheel, D. (1997). Elicitor-stimulated ion fluxes and O_2^- from the oxidative burst are essential components in triggering defense gene activation and phytoalexin synthesis in parsley. *Proceedings of the National Academy of Sciences, USA* 94: 4800-4805.

Jia, Y., McAdams, S.A., Bryan, G.T., Hershey, H.P., and Valent, B. (2000). Direct interaction of resistance gene and avirulence gene products confers rice blast resistance. *The EMBO Journal* 19: 4004-4014.

Johnson, P.R. and Ecker, J.R. (1998). The ethylene gas signal transduction pathway: A molecular perspective. *Annual Review of Genetics* 32: 227-254.

Jones, D. A., Thomas, C.M., Hammond-Kosack, K.E., Balint-Kurti, P.J., and Jones, J.D.G. (1994). Isolation of the tomato *Cf-9* gene for resistance to *Cladosporium fulvum* by transposon tagging. *Science* 266: 789-793.

Jones, O.T.G. (1994). The regulation of superoxide production by the NADPH oxidase of neutrophils and other mammalian cells. *Bioessays* 16: 919-923.

Joosten, M.H.A.J., Cozijnsen, A.J., and De Wit, P.J.G.M. (1994). Host resistance to a fungal tomato pathogen lost by a single base-pair change in an avirulence gene. *Nature* 367: 348-387.

Jorgensen, J.H. (1992). Discovery, characterization and exploitation of Mlo powdery mildew resistance in barley. *Euphytica* 63: 141-152.

Kamoun, S. (2001). Nonhost resistance to *Phytophthora*: Novel prospects for a classical problem. *Current Opinion in Plant Biology* 4: 295-300.

Kawasaki, T., Henmi, K., Ono, E., Hatakeyama, S., Iwano, M., Satoh, H., and Shimamoto, K. (1999). The small GTP-binding protein Rac is a regulator of cell death in plants. *Proceedings of the National Academy of Sciences, USA* 96: 10922-10926.

Keller, H., Pamboukdjian, N., Ponchet, M., Poupet, A., Delon, R., and Verrier, J.-L. (1999). Pathogen-induced elicitin production in transgenic tobacco generates a hypersensitive response and nonspecific disease resistance. *The Plant Cell* 11: 223-235.

Keller, T., Damude, H.G., Werner, D., Doerner, P., Dixon, R.A., and Lamb, C. (1998). A plant homolog of the neutrophil NADPH oxidase gp91phox subunit gene encodes a plasma membrane protein with Ca^{2+} binding motifs. *The Plant Cell* 10: 255-266.

Khush, R.S. and Lemaitre, B. (2000). Genes that fight infection—What the *Drosophila* genome says about animal immunity. *Trends in Genetics* 16: 442-449.

Kim, C.Y., Lee, S.H., Park, H.C., Bae, C.G., Cheong, Y.H., Choi, Y.J., Han, C.D., Lee, S.Y., Lim, C.O., and Cho, M.J. (2000). Identification of rice blast fungal elicitor-responsive genes by differential display analysis. *Molecular Plant-Microbe Interactions* 13: 470-474.

Kim, M.C., Lee, S.H., Kim, J.K., Chun, H.J., Choi, M.S., Chung, W.S., Moon, B.C., Kang, C.H., Park, C.Y., Yoo, J.H., et al. (2002). Mlo, a modulator of plant defense and cell death, is a novel calmodulin-binding protein—Isolation and characterization of a rice Mlo homologue. *Journal of Biological Chemistry* 277: 19304-19314.

Kim, M.C., Panstruga, R., Elliott, C., Muller, J., Devoto, A., Yoon, H.W., Park, H.C., Cho, M.J., and Schulze-Lefert, P. (2002). Calmodulin interacts with MLO protein to regulate defence against mildew in barley. *Nature* 416: 447-450.

Kim, Y.S., Lee, J.H., Yoon, G.M., Cho, H.S., Park, S.-W., Suh, M.C., Choi, D., Ha, H.J., Liu, J.R., and Pai, H.-S. (2000). CHRK1, a chitinase-related receptor-like kinase in tobacco. *Plant Physiology* 123: 905-915.

Kitagawa, K., Skowyra, D., Elledge, S.J., Harper, J.W., and Hieter, P. (1999). SGT1 encodes an essential component of the yeast kinetochore assembly pathway and a novel subunit of the SCF ubiquitin ligase complex. *Molecular Cell* 4: 21-33.

Klüsener, B. and Weiler, E.W. (1999). Pore-forming properties of elicitors of plant defense reactions and cellulolytic enzymes. *FEBS Letters* 459: 263-266.

Knogge, W. (1998). Fungal pathogenicity. *Current Opinion in Plant Biology* 1: 324-328.

Köhler, C., Merkle, T., and Neuhaus, G. (1999). Characterization of a novel gene family of putative cyclic nucleotide- and calmodulin-regulated ion channels in *Arabidopsis thaliana*. *The Plant Journal* 18: 97-104.

Köhler, C., Merkle, T., Roby, D., and Neuhaus, G. (2001). Developmentally regulated expression of a cyclic nucleotide-gated ion channel from *Arabidopsis* indicates its involvement in programmed cell death. *Planta* 213: 327-332.

Kooman-Gersmann, M., Honee, G., Bonnema, G., and De Wit, P.J.G.M. (1996). A high-affinity binding site for the AVR9 peptide elicitor of *Cladosporium fulvum* is present on plasma membranes of tomato and other solanaceous plants. *The Plant Cell* 8: 929-938.

Kroj, T., Rudd, J.J., Nürnberger, T., Gäbler, Y., Lee, J., and Scheel, D. (2003). Mitogen-activated protein kinases play an essential role in oxidative burst-independent expression of pathogenesis-related genes in parsley. *Journal of Biological Chemistry* 278: 2256-2264. [DOI 10.1074/jbc.M208200200]

Kunkel, B.N. and Brooks, D.M. (2002). Cross talk between signaling pathways in pathogen defense. *Current Opinion in Plant Biology* 5: 325-331.

Lauge, R., Goodwin, P.H., de Wit, P.J.G.M., and Joosten, M.H.A.J. (2000). Specific HR-associated recognition of secreted proteins from *Cladosporium fulvum* occurs in both host and non-host plants. *The Plant Journal* 23: 735-745.

Lecourieux, D., Mazars, C., Pauly, N., Ranjeva, R., and Pugin, A. (2002). Analysis and effects of cytosolic free calcium increases in response to elicitors in *Nicotiana plumbaginifolia* cells. *The Plant Cell* 14: 2627-2641.

Ligterink, W., Kroj, T., zur Nieden, U., Hirt, H., and Scheel, D. (1997). Receptor-mediated activation of a MAP kinase in pathogen defense of plants. *Science* 276: 2054-2057.

Luderer, R., Rivas, S., Nürnberger, T., Mattei, B., Van den Hooven, H.W., Van der Hoorn, R.A.L., Romeis, T., Wehrfritz, J.-M., Blume, B., Nennstiel, D., et al. (2001). No evidence for binding between resistance gene product Cf-9 of tomato and avirulence gene product AVR9 of *Cladosporium fulvum*. *Molecular Plant-Microbe Interactions* 14: 867-876.

Lyapina, S., Cope, G., Shevchenko, A., Serino, G., Tsuge, T., Zhou, C., Wolf, D.A., Wei, N., Shevchenko, A., and Deshaies, R.J. (2001). Promotion of NEDD-CUL1 conjugate cleavage by COP9 signalosome. *Science* 292: 1382-1385.

Medzhitov, R. and Janeway, C. (1997). Innate immunity: The virtues of a nonclonal system of recognition. *Cell* 91: 295-298.

Melchers, L.S. and Stuiver, M.H. (2000). Novel genes for disease-resistance breeding, *Current Opinion in Plant Biology* 3: 147-152.

Mithöfer, A., Ebel, J., Bhagwat, A.A., Boller, T., and Neuhaus-Url, G. (1999). Transgenic aequorin monitors cytosolic calcium transients in soybean cells challenged with β-glucan or chitin elicitors. *Planta* 207: 566-574.

Mithöfer, A., Fliegmann, J., Neuhaus-Url, G., Schwarz, H., and Ebel, J. (2000). The hepta-beta-glucoside elicitor-binding proteins from legumes represent a putative receptor family. *Biological Chemistry* 381: 705-713.

Navarre, D.A., Wendehenne, D., Durner, J., Noad, R., and Klessig, D.F. (2000). Nitric oxide modulates the activity of tobacco aconitase. *Plant Physiology* 122: 573-582.

Nennstiel, D., Scheel, D., and Nürnberger, T. (1998). Characterization and partial purification of an oligopeptide elicitor receptor from parsley (*Petroselinum crispum*). *FEBS Letters* 431: 405-410.

Nühse, T.S., Peck, S.C., Hirt, H., and Boller, T. (2000). Microbial elicitors induce activation and dual phosphorylation of the *Arabidopsis thaliana* MAPK 6. *Journal of Biological Chemistry* 275: 7521-7526.

Nürnberger, T. and Brunner, F. (2002). Innate immunity in plants and animals: Emerging parallels between the recognition of general elicitors and pathogen-associated molecular patterns. *Current Opinion in Plant Biology* 5: 318-324.

Nürnberger, T., Nennstiel, D., Jabs, T., Sacks, W.R., Hahlbrock, K., and Scheel, D. (1994). High affinity binding of a fungal oligopeptide elicitor to parsley plasma membranes triggers multiple defense responses. *Cell* 78: 449-460.

Nürnberger, T. and Scheel, D. (2001). Signal transmission in the plant immune response. *Trends in Plant Science* 6: 372-379.

Ono, E., Wong, H.-L., Kawasaki, T., Hasegawa, M., Kodama, O., and Shimamoto, K. (2001). Essential role of the small GTPase Rac in disease resistance of rice. *Proceedings of the National Academy of Sciences, USA* 98: 759-764.

Osman, H., Mikes, V., Milat, M.-L., Ponchet, M., Marion, D., Prange, T., Maume, B.F., Vauthrin, S., and Blein, J.-P. (2001). Fatty acids bind to the fungal elicitor cryptogein and compete with sterols. *FEBS Letters* 489: 55-58.

Osman, H., Vauthrin, S., Mikes, V., Milat, M.-L., Panabières, F., Marais, A., Brunie, S., Maume, B., Ponchet, M., and Blein, J.-P. (2001). Mediation of elicitin activity on tobacco is assumed by elicitin-sterol complexes. *Molecular Biology of the Cell* 12: 2825-2834.

Peart, J.R., Lu, R., Sadanandom, A., Malcuit, I., Moffett, P., Brice, D.C., Schauser, L., Jaggard, D.A.W., Xiao, S., Coleman, M.J., et al. (2002). Ubiquitin ligae-associated protein SGT1 is required for host and nonhost disease resistance in plants. *Proceedings of the National Academy of Sciences, USA* 99: 10865-10869.

Petersen, M., Brodersen, P., Naested, H., Andreasson, E., Lindhart, U., Johansen, B., Nielsen, H.B., Lacy, M., Austin, M.J., Parker, J.E., et al. (2000). *Arabidopsis* MAP kinase 4 negatively regulates systemic acquired resistance. *Cell* 103: 1111-1120.

Piedras, P., Rivas, S., Dröge, S., Hillmer, S., and Jones, J.D.G. (2000). Functional, c-myc-tagged Cf-9 resistance gene products are plasma-membrane localized and glycosylated. *The Plant Journal* 21: 529-536.

Qutob, D., Kamoun, S., Tyler, B.M., and Gijzen, M. (2002). Expression of a *Phytophthora sojae* necrosis-inducing protein occurs during transition from biotrophy to necrotrophy. *The Plant Journal* 32: 361-373.

Ramonell, K., Zhang, B., Ewing, R., Chen, Y., Xu, D., Stacey, G., and Somerville, S. (2002). Microarray analysis of chitin elicitation in *Arabidopsis thaliana*. *Molecular Plant Pathology* 3: 301-311.

Ren, D., Yang, H., and Zhang, S. (2002). Cell death mediated by MAPK is associated with hydrogen peroxide production in *Arabidopsis*. *Journal of Biological Chemistry* 277: 559-565.

Rivas, S., Romeis, T., and Jones, J.D. (2002). The Cf-9 disease resistance protein is present in an approximately 420-kilodalton heteromultimeric membrane-associated complex at one molecule per complex. *The Plant Cell* 14: 689-702.

Robson, C.A. and Vanleberghe, G.C. (2002). Transgenic plant cells lacking mitochondrial alternative oxidase have increased susceptibility to mitochondria-dependent and independent pathways of programmed cell death. *Plant Physiology* 129: 1908-1920.

Rohe, M., Gierlich, A., Hermann, H., Hahn, M., Schmidt, B., Rosahl, S., and Knogge, W. (1995). The race-specific elicitor, NIP1, from the barley pathogen, *Rhynchosporium secalis,* determines avirulence on host plants of the Rrs1 resistance genotype. *The EMBO Journal* 14: 4168-4177.

Romeis, T., Ludwig, A.A., Martin, R., and Jones, J.D.G. (2001). Calcium-dependent protein kinases play an essential role in a plant defence response. *The EMBO Journal* 20: 5556-5567.

Romeis, T., Piedras, P., and Jones, J.D.G. (2000). Resistance gene-dependent activation of a calcium-dependent protein kinase in the plant defense response. *The Plant Cell* 12: 803-815.

Romeis, T., Piedras, P., Zhang, S., Klessing, D.F., Hin, H., and Jones, J.D. (1999). Rapid Avr9- and Cf-9-dependent activation of MAP kinases in tobacco cell cultures and leaves: Convergence of resistance gene, elicitor, wound, and salicylate responses., *The Plant Cell* 11: 273-287.

Sagi, M. and Fluhr, R. (2001). Superoxide production by plant homologues of the gp91[phox] NADPH oxidase: Modulation of activity by calcium and by tobacco mosaic virus infection. *Plant Physiology* 126: 1281-1290.

Scheel, D. (1998). Resistance response physiology and signal transduction. *Current Opinion in Plant Biology* 1: 305-310.

Scheel, D. (2002a). Oxidative burst and the role of reactive oxygen species in plant-pathogen interactions. In Inzé, D. and Van Montagu, M. (Eds.), *Oxidative Stress in Plants* (pp. 137-153). London, New York: Taylor and Francis.

Scheel, D. (2002b). Signal transduction elements. In Oksman-Caldentey, K.-M. and Barz, H.W. (Eds.), *Plant Biotechnology and Transgenic Plants* (pp. 427-444). New York: Marcel Dekker, Inc.

Schenk, P.M., Kazan, K., Wilson, I., Anderson, J.P., Richmond, T., Somerville, S.C., and Manners, J.M. (2000). Coordinated plant defense responses in *Arabidopsis* revealed by microarray analysis. *Proceedings of the National Academy of Sciences, USA* 97: 11655-11660.

Schwechheimer, C., Serino, G., Callis, J., Crosby, W.L., Lyapina, S., Deshaies, R.J., Gray, W.M., Estelle, M., and Deng, X.-W. (2001). Interactions of the COP9 signalosome with the E3 ubiquitin ligase SCF[TIR1] in mediating auxin response. *Science* 292: 1379-1382.

Seo, H.S., Song, J.T., Cheong, J.-J., Lee, Y.-H., Lee, Y.-W., Hwang, I., Lee, J.S., and Choi, Y.D. (2001). Jasmonic acid carboxyl methyltransferase: A key enzyme for jasmonate-regulated plant responses. *Proceedings of the National Academy of Sciences, USA* 98: 4788-4793.

Shirasu, K., Lahaye, T., Tan, M-W., Zhou, F., Azevedo, C., and Schulze-Lefert, P. (1999). A novel class of eukaryotic zinc-binding proteins is required for disease resistance signaling in barley and development in *C. elegans*. *Cell* 99: 355-366.

Simon-Plas, F., Elmayan, T., and Blein, J.P. (2002). The plasma membrane oxidase NtrbohD is responsible for AOS production in elicited tobacco cells. *The Plant Journal* 31: 137-147.

Suharsono, U., Fujisawa, Y., Kawasaki, T., Iwasaki, Y., Satoh, H., and Shimamoto, K. (2002). The heterotrimeric G protein alpha subunit acts upstream of the small GTPase Rac in disease resistance of rice. *Proceedings of the National Academy of Sciences, USA* 99: 13307-13312.

Takken, F.L.W. and Joosten, M.H.A.J. (2000). Plant resistance genes: Their structure, function and evolution. *European Journal of Plant Pathology* 106: 699-713.

Tang, X., Xie, M., Kim, Y.J., Zhou, J., Klessig, D.F., and Martin, G.B. (1999). Overexpression of *Pto* activates defense responses and confers broad resistance. *The Plant Cell* 11: 15-30.

Thoma-Uszynski, S., Stenger, S., Takeuchi, O., Ochoa, M.T., Engele, M., Sieling, P.A., Barnes, P.F., Röllinghoff, M., Bölcskei, P.L., Wagner, M., et al. (2001). Induction of direct antimicrobial activity through mammalian Toll-like receptors. *Science* 291: 1544-1547.

Thomma, B.P., Cammue, B.P., and Thevissen, K. (2002). Plant defensins. *Planta* 216: 193-202.

Tör, M., Gordon, P., Cuzick, A., Eulgem, T., Sinapidou, E., Mert-Türk, F., Can, C., Dangl, J.L., and Holub, E.B. (2002). *Arabidopsis* SGT1b is required for defense signaling conferred by several downy mildew resistance genes. *The Plant Cell* 14: 993-1003.

Torres, M.A., Dangl, J.L., and Jones, J.D.G. (2002). *Arabidopsis* gp91[phox] homologues AtrbohD and AtrbohF are required for accumulation of reactive oxygen intermediates in the plant defense response. *Proceedings of the National Academy of Sciences, USA* 99: 517-522.

Torres, M.A., Onouchi, H., Hamada, S., Machida, C., Hammond-Kosack, K.E., and Jones, J.D.G. (1998). Six *Arabidopsis thaliana* homologues of the human respiratory burst oxidase (*gp91[phox]*). *The Plant Journal* 14: 365-370.

Tyler, B.M. (2002). Molecular basis of recognition between phytophthora pathogens and their hosts. *Annual Review of Phytopathology* 40: 137-167.

Umemoto, N., Kakitani, M., Iwamatsu, A., Yoshikawa, M., Yamaoka, N., and Ishida, I. (1997). The structure and function of a soybean beta-glucan-elicitor-binding protein. *Proceedings of the National Academy of Sciences, USA* 94: 1029-1034.

Underhill, D.M. and Ozinsky, A. (2002). Toll-like receptors: Key mediators of microbe detection. *Current Opinion in Immunology* 14: 103-110.

van der Biezen, E.A. and Jones, J.D.G. (1998). Plant disease resistance proteins and the gene-for-gene concept. *Trends in Biochemical Sciences* 23: 454-456.

Van der Hoorn, R.A., De Wit, P.J., and Joosten, M.H. (2002). Balancing selection favors guarding resistance proteins. *Trends in Plant Science* 7: 67-71.

Veit, S., Wörle, J.M., Nürnberger, T., Koch, W., and Seitz, H.U. (2001). A novel protein elicitor (PaNie) from *Pythium aphanidermatum* induces multiple defense responses in carrot, *Arabidopsis,* and tobacco. *Plant Physiology* 127: 832-841.

Verberne, M.C., Verpoorte, R., Bol, J.F., Mercado-Blanco, J., and Linthorst, H.J.M. (2000). Overproduction of salicylic acid in plants by bacterial transgenes enhances pathogen resistance. *Nature Biotechnology* 18: 779-783.

Wendehenne, D., Pugin, A., Klessig, D.F., and Durner, J. (2001). Nitric oxide: Comparative synthesis and signaling in animal and plant cells. *Trends in Plant Science* 6: 177-183.

Wevelsiep, L., Rüpping, E., and Knogge, W. (1993). Stimulation of barley plasmalemma H^+-ATPase by phytotoxic peptides from the fungal pathogen *Rhynchosporium secalis*. *Plant Physiology* 101: 297-301.

Xing, T., Wang, X.J., Malik, K., and Miki, B.L. (2001). Ectopic expression of an *Arabidopsis* calmodulin-like domain protein kinase-enhanced NADPH oxidase activity and oxidative burst in tomato protoplasts. *Molecular Plant-Microbe Interactions* 14: 1261-1264.

Xu, H. and Heath, M.C. (1998). Role of calcium in signal transduction during the hypersensitive response caused by basidiospore-derived infection of the cowpea rust fungus. *The Plant Cell* 10: 585-597.

Yang, K.Y., Liu, Y., and Zhang, S. (2001). Activation of a mitogen-activated protein kinase pathway is involved in disease resistance in tobacco. *Proceedings of the National Academy of Sciences, USA* 98: 741-746.

Yeh, E.T., Gong, L., and Kamitani, T. (2000). Ubiquitin-like proteins: New wines in new bottles. *Gene* 248: 1-14.

Zhang, B., Ramonell, K., Somerville, S., and Stacey, G. (2002). Characterization of early, chitin-induced gene expression in *Arabidopsis*. *Molecular Plant-Microbe Interactions* 15: 963-970.

Zhang, S. and Klessig, D.F. (1998). The tobacco wounding-activated mitogen-activated protein kinase is encoded by *SIPK*. *Proceedings of the National Academy of Sciences, USA* 95: 7225-7230.

Zhang, S. and Klessig, D.F. (2001). MAPK cascades in plant defense signaling. *Trends in Plant Science* 6: 520-527.

Zhang, S. and Liu, Y. (2001). Activation of salicylic acid-induced protein kinase, a mitogen-activated protein kinase, induces multiple defense responses in tobacco. *The Plant Cell* 13: 1877-1889.

Zimmermann, S., Nürnberger, T., Frachisse, J.-M., Wirtz, W., Guern, J., Hedrich, R., and Scheel, D. (1997). Receptor-mediated activation of a plant Ca^{2+}-permeable ion channel involved in pathogen defense. *Proceedings of the National Academy of Sciences, USA* 94: 2751-2755.

Zinecker, H. (2001). Reaktive Sauerstoffspezies in der pflanzlichen Pathogenabwehr: Isolierung und Charakterisierung von Genen aus *Petroselinum crispum* L., die für putative NADPH-Oxidasen kodieren. PhD thesis, Martin-Luther-Universität Halle Wittenberg, Halle (Saale).

Chapter 2

Cellular Expression of Resistance to Fungal Plant Pathogens

Denny G. Mellersh
Michèle C. Heath

INTRODUCTION

Recently a great deal of importance has been placed on the eluci-dation of plant resistance mechanisms through the use of biochemical and genetic techniques. In particular, the development of *Arabidopsis* as a model system has led to the rapid isolation of numerous plant genes that provide resistance to fungal pathogens (Adam and Somer-ville, 1996; Xiao et al., 2001). With some notable exceptions (Eul-gem et al., 1999), most recent molecular biological studies have fo-cused on global leaf responses that occur in association with plant resistance to fungi. However, many cellular events associated with plant resistance are transient or are specific to certain plant cells or to localized areas of individual cells. This is especially true in the case of plant resistance to fungi that directly penetrate plant cell walls as an integral part of their infection process. In these situations, bio-chemical and genetic studies are most useful when combined with detailed cytological work (e.g., Adam and Somerville, 1996).

Biotrophic fungi that directly invade living epidermal cells provide the best opportunity to observe events associated with resistance at the cellular level because (1) events in epidermal cells can be directly viewed by light microscopy, and (2) biotrophic fungi do not kill the susceptible plant cells they invade, thus allowing plant cell responses to be observed during wall penetration and intracellular fungal growth. Among the biotrophic fungal plant pathogens, the most detailed cyto-logical studies have been performed on plant interactions with the rust and powdery mildew fungi, and these interactions will therefore be the primary focus of this review. Rust fungi typically have a dikaryotic parasitic stage in which the fungus enters plant tissue via stomata, grows intercellularly, and only penetrates plant cells to form feeding structures (haustoria). However, they generally also have a second, monokaryotic, parasitic stage which develops from basidiospores and which, prior to forming an intercellular mycelium with haustoria, penetrates the plant directly through epidermal cells. Powdery mildew fungi typically grow on the plant surface, with only feeding structures (haustoria) penetrating underlying epidermal cells. A diagrammatic representation of typical epidermal infection processes by these two fungal taxa is shown in Figure 2.1.

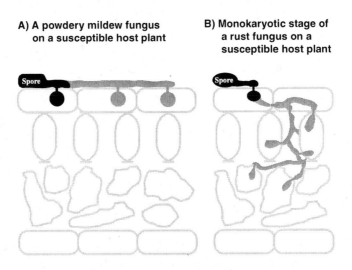

FIGURE 2.1. Typical infection process of two types of biotrophic fungi on susceptible host plants. (A) A powdery mildew spore germinates on the surface of a leaf to form an appressorium prior to directly penetrating and forming a feeding structure called a haustorium (all shown in black) within an underlying epidermal cell. The fungus then continues to grow (shown in dark gray) on the surface of the leaf forming additional haustoria in adjacent epidermal cells. (B) A rust fungal basidiospore germinates on the surface of a leaf to form an appressorium prior to directly penetrating and forming a feeding structure called an intraepidermal vesicle (all shown in black) within an underlying epidermal cell. The fungus then continues to grow (shown in dark gray) by tip growth, entering adjacent epidermal cells prior to growing into the intercellular spaces of the leaf and, finally, producing monokaryotic haustoria within mesophyll cells.

BACKGROUND

Potentially Defensive Plant Cell Wall-Associated Responses to Fungi

Although fungal pathogens are normally successful in penetrating the cell walls of their host species, attempted penetration in nonhost plants frequently results in penetration failure. Constitutive plant features can play a role in restricting fungal growth at this stage of infection; however, penetration failure is often dependent on active plant cell metabolism, and chemical interference with the latter process results in increased penetration success (Sherwood and Vance, 1980). Numerous wall-associated responses and modifications, often localized to the site in the cell wall where the fungus is penetrating, have been reported that could theoretically hinder fungal progress through the cell wall (Figure 2.2A). These include the extracellular generation

A) Penetration-induced wall-associated responses

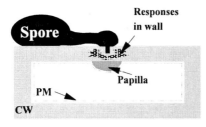

B) Hypersensitive cell death following successful cell penetration

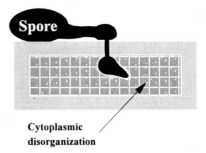

FIGURE 2.2. Potentially defensive cellular responses associated with attempted and successful fungal penetration of plant cells. (A) Attempted (or in some cases even successful) fungal penetration of the plant cell wall elicits a variety of responses in the plant cell wall (CW) as well as the deposition of callose-containing papillae between the CW and plasma membrane (PM). (B) Successful fungal penetration frequently results in the rapid death of the infected cell (hypersensitive cell death) as characterized by the presence of cytoplasmic disorganization. For (A) and (B), fungal structures are shown in black.

of reactive oxygen species (ROS), such as hydrogen peroxide (H_2O_2) and superoxide (O_2^-) (Thordal-Christensen et al., 1997; Mellersh et al., 2002), impregnation of the cell wall with phenolics or silica (Heath and Stumpf, 1986; Aist and Bushnell, 1991), oxidative cross-linking of preexisting or induced cell wall phenolics or proteins (Aist and Bushnell, 1991; Thordal-Christensen et al., 1997), and the deposition of heterogeneous callose-containing papillae or wall appositions between the plant cell wall and plasma membrane (Aist, 1976).

What Triggers Localized Plant Cell Responses During Cell Wall Penetration?

Current data suggest that plant responses to the penetration process are primarily the result of perception of plant cell wall components released during localized wall degradation (Heath et al., 1997). Localized application of hemicellulase (a mixture of wall-degrading enzymes) to small scratches in the cuticle of cowpea epidermal cells results, sequentially, in the localized extracellular generation of H_2O_2, accumulation of phenolic compounds, and finally cell wall protein cross-linking (Mellersh et al., 2002). Interestingly, insertion of a glass microneedle into a cell will result in deposition of callose (a β-1,3-linked glucan) around its tip, but not in association with accumulation of phenolic compounds in the cell wall (Russo and Bushnell, 1989). This suggests that plant cells respond differentially to physical versus chemical cues associated with cell wall penetration. However, a comparison of wound plugs formed in response to insertion of a glass microneedle with papillae formed in response to the barley powdery mildew fungus revealed that only wound plugs contained cellulose and pectin, and only fungal-induced papillae contained phenols and basic staining material (Russo and Bushnell, 1989). These observations suggest that fungal penetration results in different, rather than simply additional, induction cues than does physical intrusion.

Detailed cytological studies indicate that while certain plant responses to fungal penetration (namely extracellular H_2O_2 generation and accumulation of phenolic compounds) are seen during penetration of nonhosts by both rust fungi and powdery mildew fungi (Mellersh et al., 2002), powdery mildew fungi also elicit the addi-

tional generation of extracellular superoxide. This further supports
the hypothesis that while certain unavoidable events associated with
wall penetration elicit predictable plant responses, these responses
are also influenced by other, as yet undetermined, activities of the
penetrating fungus.

Postpenetration Responses

Plant cell responses to the presence of parasitic fungi within the
cell lumen have been studied by light or electron microscopy for almost a century. As might be expected for biotrophic pathogens that
enter living plant cells, subcellular rearrangements, such as changes
in the plant's endomembrane system and the cytoskeleton, occur in
invaded cells of both genetically resistant and susceptible plants
(Heath and Škalamera, 1997). In susceptible plants, the invaded cell
remains alive despite the presence of the fungus which, although inside the cell, remains separated from the plant cytoplasm by the plant
plasma membrane. In disease-resistant plants, however, fungal growth
is eventually restricted, sometimes so rapidly that the pathogen does
not get beyond the first-invaded plant cell. Although the cessation of
fungal growth within a cell may be associated with encasement of the
pathogen in callose and the plant cell remains alive (Škalamera et al.,
1997), the most common expression of resistance is a rapid death of
the invaded cell (Figure 2.2B). This hypersensitive response (HR)
was first recognized by Stakman (1915), and was first extensively
studied in living cells using potato and the oomycete pathogen, *Phytophthora infestans* (Tomiyama, 1960). Experiments using this system first demonstrated that the hypersensitive response requires energy in its execution (Tomiyama, 1967) and is, therefore, a form of
programmed cell death.

Potential versus Actual Defense Mechanisms

In any given plant-fungal interaction, a number of potentially inhibitory plant responses may be present at the same time. Each type
of response, however, may not contribute equally, if at all, to penetration resistance. Unequivocal evidence as to which response(s) is most
important is scarce, because some potentially inhibitory compounds

associated with penetration resistance may not actually be responsible for stopping fungal growth (Perumalla and Heath, 1989; McLusky et al., 1999). In addition, removal of one effective defense response can in some cases lead to compensation by another effective response (Heath and Stumpf, 1986). Complicating matters further, some plant responses such as the accumulation of small molecular weight antimicrobial compounds known as phytoalexins (Hammerschmidt, 1999) are undetectable at the level of the light microscope, despite the potential importance of their cellular and subcellular localization. Only a few phytoalexins have been studied in living plant cells. These include a phytoalexin from broad bean *(Vicia faba)* leaves, which was detected in cell vacuoles by fluorescence under ultraviolet light (Mansfield et al., 1974) and the pigmented phytoalexin in sorghum *(Sorghum bicolor)* (Snyder and Nicholson, 1990). The latter authors demonstrated the localized accumulation of these red phytoalexins in membrane-bound inclusion bodies that migrated to the attempted penetration site of *Colletotrichum graminicola*. Colorless inclusions coalesced with one another and gradually darkened in intensity, demonstrating that the phytoalexins were produced within the inclusions themselves. The inclusions eventually deposited their contents to the cytoplasm and were also seen to accumulate in the overlying fungus. Such examples are rare, however, and although it is tempting to focus on those responses that are most easily microscopically identified, researchers must be ever mindful that other responses, perhaps unique to certain plant-fungal interactions, may also play prominent roles.

DETAILED REVIEW

Resistance Expressed During Cell Wall Penetration

The Role of Reactive Oxygen Species (ROS)

The localized, cytologically detectable, extracellular generation of ROS has been reported to be associated with penetration events in a number of plant-fungal interactions (Thordal-Christensen et al., 1997; Mellersh et al., 2002). The generation of hydrogen peroxide (H_2O_2) in particular, is often correlated with penetration failure (Hückelhoven, Trujillo, and Kogel, 2000; Hückelhoven, Dechert, and Kogel,

2001) and is apparently at least partially responsible for the latter in nonhost plants (Mellersh et al., 2002) since its enzymatic removal allows increased penetration success of rust and powdery mildew fungi on nonhost plants (Mellersh et al., 2002). The various potential enzymatic sources of H_2O_2 is a topic of considerable recent debate and may in fact vary depending on the particular plant-pathogen interaction (Bolwell, 1999). Regardless of its enzymatic source, H_2O_2 been suggested to participate in plant defense responses in a variety of ways, including by acting as a signaling agent, or via oxidative changes to fortify plant cell walls. In addition, H_2O_2 may be directly toxic to fungi since exogenously added or enzymatically generated H_2O_2 is toxic to fungi in vitro (Lu and Higgins, 1999) and in tobacco leaf disks (Peng and Kuc, 1992). In the nonhost resistance of cowpea to the plantain powdery mildew fungus, *Erysiphe cichoracearum,* however, H_2O_2 appears to reduce fungal penetration success indirectly by either (1) acting as a signaling molecule for new gene expression, resulting in the secretion of other penetration-inhibiting molecules, or (2) interacting with such molecules in the wall in such a way as to render them inhibitory to penetration (Mellersh et al., 2002). Although the inhibitory molecules in the aforementioned study were not clearly identified, data suggest that these may be phenolic in nature and that cytoplasmic activity was directly or indirectly needed for their secretion into the plant wall. It is interesting to note, however, that in the case of resistant young tomato leaves inoculated with *Colletotrichum coccodes,* H_2O_2 generation within the wall did not require involvement of the cell protoplast and, alone, seemed both necessary and sufficient for penetration failure (Mellersh et al., 2002). The difference between the role of H_2O_2 in these different systems is a good example of why it is often difficult to generalize about defense responses at the cellular level.

Phenolic Compounds

Phenolic compounds frequently accumulate locally in response to fungal infection (Nicholson and Hammerschmidt, 1992) and are potentially involved in penetration resistance. Phenolic compounds may be particularly important in resistance to powdery mildew fungi, since inhibition of autofluorescent phenolic compound accumulation using phenylalanine ammonia-lyase (PAL) inhibitors led to increased penetration success of *Erysiphe graminis* on both young (Carver et al., 1992) susceptible oat (*Avena sativa* L.) plants and on oat plants expressing adult plant resistance (Carver et al., 1996). Phenolic accumulation in leaves is frequently detected by their tendency to autofluoresce under blue or UV light. This method of detection, however, does not discriminate among a wide variety of chemically distinct phenolic compounds that may accumulate during plant-fungal interactions. Because the type of phenolics present may be an important determinant of penetration failure, recent studies have focused on identifying the individual chemical species present in various interactions. Feruloyl-3' methoxytyramine specifically accumulates in granular inclusions in onion (*Allium sativa* L.) epidermal cells at attempted penetration sites of *Botrytis allii* (McLusky et al., 1999), and although this compound was not antimicrobial in vitro, it cannot be ruled out that these phenolics may provide some sort of physical barrier to fungal penetration. In one study, von Röpenack and colleagues (1998) demonstrated that a single phenolic compound, *p*-coumaroylhydroxyagmatine (*p*-CHA), differentially accumulated in barley plants expressing *mlo*-mediated penetration resistance, and that *p*-CHA was antifungal both in vitro and in vivo. However, it remains unclear whether *p*-CHA is any more important than a variety of other responses, such as potentially penetration-inhibitory changes in cell wall-associated papillae (reflected by the latter's increased resistance to alkaline degradation), which are also associated with *mlo*-mediated penetration resistance.

Papillae

Cell wall-penetrating fungi frequently trigger the localized formation of heterogeneous deposits between the plant plasma membrane and cell wall that are referred to as cell wall appositions or papillae. With a few exceptions (Aist, 1976), papillae are not usually found unless the fungal pathogen has managed to breach the plant cell wall, suggesting this is primarily a response to localized physical trauma. Regardless, papillae appear to play an important role in preventing successful fungal penetration in a wide variety of plant species (Sherwood and Vance, 1980). Nevertheless, despite the almost ubiquitous nature of the papilla response, papillae are not always an effective barrier to fungal ingress, and papillae are commonly associated with *successful* penetration by powdery mildew fungi (Aist and Bushnell, 1991). In these latter cases, the relative timing of papilla formation during the penetration process may be an important determinant of whether papillae are effective (Aist and Israel, 1977).

Papillae could potentially present either a physical or chemical barrier (or both) to fungal ingress, and the aspects of papillae that are responsible for stopping the fungus have been the subject of considerable study. Papillae frequently are impregnated with impermeable compounds such as silica (Carver et. al., 1987), and this could potentially provide a barrier to the exchange of nutrients or molecules involved in the plant-pathogen "dialogue" (Smart et al., 1987), as well as a structural barrier against fungal penetration. Silica has been implicated in the failure of rust fungi to form haustoria in nonhost plants (Stumpf and Heath, 1985; Perumalla and Heath, 1991); however, no similar data demonstrate a causative role in penetration resistance to powdery mildew fungi (Carver et al., 1987). Potentially toxic molecules can also accumulate in papillae (von Röpenack et al., 1998; Mellersh et al., 2002), and these molecules may be similar to those seen in modified cell walls. Just as unmodified cell walls seldom present an obstacle to direct-penetrating fungi in their host species, callose—the main structural component of many papillae—is not likely to be particularly relevant itself in lending penetration resistance (Smart et al., 1986; Perumalla and Heath, 1989). We assume, therefore, that it is both the structural and chemical changes associated with papilla formation that are more important than the papilla itself in stopping fungal growth.

Cytoplasmic Rearrangements That Accompany Wall-Associated Responses

A variety of changes are induced in the cytoplasm of plant cells during early stages of the penetration process in association with resistance. Redirection of cytoplasm toward the penetration point is one of the earliest of these responses and is associated with penetration failure of *Erysiphe graminis* and *Erysiphe pisi,* a pathogen and nonpathogen, respectively, on barley (Kobayashi et al., 1993). It is believed that cytoplasm redirected to the penetration point contains necessary components required to alter the plant cell wall, rendering it resistant to fungal penetration (Aist and Bushnell, 1991).

Cytoskeletal elements may be involved in the direction of signals or may redirect the raw materials required to mount localized wall-associated responses. Rearrangements of the actin cytoskeleton seem particularly important in regard to penetration arrest (McLusky et al., 1999) and disruption of microfilaments causes the loss of localized wall-associated responses (Mellersh et al., 2002) and allows numerous fungi to gain access to the cells of otherwise resistant nonhost plants (Kobayashi et al., 1997).

Plant nuclei migrate to sites of both localized mechanical (Gus-Mayer et al., 1998) and chemical (Heath et al., 1997) perturbation, and association of the plant cell nucleus with the penetration point, and in some cases with the developing fungus inside penetrated cells, has been reported for a number of plant-fungal interactions (Heath and Škalamera, 1997). Although the importance of this association has not been demonstrated, it is possible that the close proximity of the plant cell nucleus with the penetration point reflects a need for a localized source of mRNA for protein synthesis that might be required for the expression of localized defensive responses. Supporting this, the nucleus does not remain in constant association with the penetrating fungus in interactions between rust fungi and their host plants, where cytologically detectable plant responses are absent (Heath et al., 1997). However, in nonhost plants in which rust fungi trigger a variety of cytologically detectable responses (Mellersh et al., 2002), the nucleus is in more or less constant association with the fungus.

Dynamic cellular processes appear to be required in the vast majority of cases in which coordination of a rapid localized plant defense response occurs. What is more surprising perhaps is the fact that, as described earlier for extracellular H_2O_2 generation, some plant cells appear to mount similar effective responses without any participation from the cell protoplast (Mellersh et al., 2002). Future research will hopefully lead us to understand both the reasons behind these major differences and the underlying mechanistic bases behind them.

Resistance Expressed Following Successful Penetration: The Hypersensitive Response

The Hypersensitive Response and Hypersensitive Cell Death

Although penetration failure is common in nonhost interactions, penetration is usually successful when fungi attempt to penetrate the cells of their host species. In the case of resistant host genotypes, resistance is often manifested by the hypersensitive death of the invaded cell in association with cessation of fungal growth. The hypersensitive response involves the endogenously programmed death of the plant cell and the induction of defenses, such as the modification of cell walls and accumulation of potential antimicrobial compounds (Heath, 2000b) (see Chapter 3). These latter responses, which often involve the participation of neighboring cells, are likely to be important in restricting the growth of nonbiotrophic pathogens for which the death of the invaded cell is not necessarily a liability. For biotrophic fungi that require living plant cells to survive, the potential relationship between cell death and resistance is more obvious, and for this reason it is often helpful to conceptually separate the hypersensitive cell death from the hypersensitive response.

Morphological Features of Hypersensitive Cell Death

Although host resistance to biotrophic fungi can be manifested by the death of either individual cells or larger numbers of cells, the most detailed cytological studies have focused on those systems in which only the initially infected epidermal cell dies while those surrounding it remain alive. At least in biotrophic interactions, cell death does not usually occur until the fungus has entered the cell lumen, and observation of the events preceding cell death in living rust-infected epidermal cells has demonstrated a number of cytological features of the death process that predictably occur following formation of an intraepidermal vesicle in resistant host cells. One of the earliest signs of the onset of cell death in the rust interaction is a change in the appearance of the plant cell nucleus (Heath et al., 1997). This is followed by the cessation of cytoplasmic streaming (possibly associated with the loss of cortical microtubules) (Škalamera and Heath, 1998), the appearance of particles exhibiting Brownian motion within the vacuole, and eventually protoplast collapse (Chen and Heath, 1991).

Host resistance to rust fungi and powdery mildew fungi alike is mediated by parasite-specific recognition events. Interestingly, the cytological events that occur in host resistance responses to both fungi are remarkably similar (Hazen and Bushnell, 1983; Chen and Heath, 1991). In contrast, hypersensitive cell death in nonhost plants is not necessarily under the control of parasite-specific recognition (Heath, 2000b) and is morphologically distinct from cell death in the former two interactions (Christopher-Kozjan and Heath, unpublished data). Likewise, the hypersensitive cell death that occurs in response to the oomycete *Phytopthora infestans* occurs in both resistant and susceptible plant cells (i.e., regardless of the presence or absence of parasite-specific recognition) and is visually quite different from interactions involving specific recognition (Freytag et al., 1994). The significance of these morphological differences is largely unknown; however, recent evidence indicates that the cell death process that occurs as a result of specific recognition events differs mechanistically from that involving nonspecific recognition (Christopher-Kozjan and Heath, unpublished data).

The Role of Calcium

Detailed cytochemical and stereological electron microscopic studies of the various stages of host cell death in response to rust fungal infection have demonstrated that resistance-specific recognition events occur while the fungus is still penetrating the epidermal wall (Xu and Heath, 1998; Mould and Heath, 1999). Rust fungi do not trigger many of the wall-associated responses normally elicited during the penetration process by most fungi (Mellersh and Heath, 2001), and as such provide a unique opportunity to study the intimate details of hypersensitive cell death without interference from extraneous events unassociated with the death process itself. Confocal microscopy studies using calcium reporter dyes microinjected into living epidermal cells at various stages of rust fungal invasion show that a prolonged, but transient, increase in cytosolic calcium levels occurs while the fungus is penetrating the epidermal walls of resistant host cells (Xu and Heath, 1998) prior to fungal contact with the plant plasma membrane. Pharmacological studies have demonstrated that this increase in cytosolic calcium levels primarily originates from extracellular stores, is required for induction of cell death to occur, and is absent in susceptible cells at the same stage of infection.

Requirement for Transcription and Translation

Detailed stereological studies investigating the number of nuclear pores on the nuclear envelope indicative of levels of nuclear to cytoplasmic trafficking, and nucleolar granularity indicative of relative levels of transcriptional activity, demonstrate that resistance-specific increases in transcription and nuclear trafficking accompany the increases in cytosolic calcium levels in rust-infected resistant cells (Mould and Heath, 1999). Interestingly, a localized increase in translation indicated by polyribosome density on the endoplasmic reticulum was associated with the area immediately beneath the penetration point. Moreover, heat shock or pharmacological treatments that interfered with plant cell protein translation inhibited cell death in response to both rust fungi (Heath et al., 1997) and powdery mildew fungi (Hazen and Bushnell, 1983). The transcription of a number of genes potentially involved in cell death was shown to be preferentially up- or down-regulated in barley during the hypersensitive response to *E. graminis* (Hückelhoven, Dechert, Trujillo, and Kogel, 2001). However, these genes were also differentially expressed to some extent in cells expressing penetration resistance, and genes with roles in defense aside from hypersensitive cell death are also up- or down-regulated following powdery mildew infection (Gregersen et al., 1997).

Reactive Oxygen Species

Reactive oxygen species have been found to be associated with hypersensitive cell death in a variety of plant pathosystems. Cytologically detectable accumulation of H_2O_2 (Thordal-Christensen et al., 1997) and O_2 (Hückelhoven and Kogel, 1998) both occur during the hypersensitive response of barley to the barley powdery mildew fungus but are absent in mutant *Mla* or *Rar* plants that do not undergo a hypersensitive response (Hückelhoven, Fodor, et al., 2000). Importantly, however, barley plants expressing *Mlg*-mediated resistance are associated with a successful papilla response and epidermal cell death in the absence of detectable O_2^- (Hückelhoven and Kogel, 1998), suggesting the O_2^- is not necessarily a prerequisite of cell death in this system. Scavengers of reactive oxygen species reduced the number of autofluorescent cells indicative of cell death in resistant cowpea cells responding to the cowpea rust fungus (Chen and Heath, 1994). However, a more recent study demonstrated that ROS do not accumulate in this system until later in the death process, and scavengers had no effect on cytoplasmic disorganization, a more reliable marker of cell death (Heath, 1998). These latter results suggest that the role of reactive oxygen species may relate more to the accumulation of autofluorescent phenolic compounds that are deposited in dying cells by living adjacent cells than to the process of cell death itself. The involvement of reactive oxygen species in hypersensitive cell death remains particularly enigmatic, especially given that even small, perhaps visually undetectable, amounts of the former may interact with other signaling molecules such as salicylic acid (Shirasu et al., 1997) and nitric oxide (Delledonne et al., 2001) in amplification of death-related signals.

How Do Fungi Cope with Defensive Cellular Plant Responses to Successfully Infect Their Host Plants?

Inducible plant defense responses appear to be the almost inevitable result of attempted fungal penetration, and as a result, any given plant pathogen seldom infects more than a small number of plant species. Nevertheless, fungi successfully infect the tissues of their host plants, and this must reflect the ability of these fungi to deal with adverse plant responses in some way.

Powdery Mildew Fungi and the Basis of Induced Accessibility

Attempted penetration of nonhost plant cells by powdery mildew fungi results in a high frequency of penetration failure. Interestingly, however, successful penetration of host barley cells by *E. graminis* can lead to increased penetration success by the nonpathogen of barley *(E. pisi)* when the latter is inoculated subsequently to the same (Kunoh et al., 1985) or adjacent (Kunoh et al., 1986) epidermal cells, a phenomenon known as induced accessibility. Importantly, this effect is dependent on the formation of haustoria by *E. hordei* (Yamaoka et al., 1994) and by the dikaryotic stage of the cowpea rust fungus, *Uromyces vignae* (Heath, 1983). For the powdery mildew system, induced accessibility may be a result of nutrient (i.e., sugar) depletion in the infected host cell (Yamaoka et al., 1994), since accessibility is reversed in the presence of exogenous glucose. This may not, however, be true for the rust system, where fungal haustoria are in energy-rich mesophyll cells rather than chloroplast-lacking epidermal cells. The effect is also seen in susceptible host interactions, where successful attacks by *E. graminis* on barley and oats renders attacked cells more susceptible to a subsequent attack by the same fungus (Lyngkjaer and Carver, 2000). This, the authors suggest, will be of great consequence in the field where plants that are initially fundamentally susceptible will become increasingly more susceptible with subsequent cycles of attack.

One of the primary challenges of biotrophy must be the need to keep the invaded plant cell alive. Despite the fact that cell death almost inevitably follows fungal penetration of nonhost plant cells (Fernandez and Heath, 1989), biotroph-infected cells frequently remain alive for days or even weeks in susceptible host plants. Successful formation of haustoria in host cells by *E. graminis* completely suppresses the host-nonspecific cell death response to later attacks by the same fungus (Carver et al., 1999), and importantly, infection by a virulent race suppresses host-specific resistance mediated by recognition of the *Mla1* resistance gene in barley (Lyngkjaer et al., 2001).

Rust Fungi and the Lack of Wall-Associated Responses to the Penetration Process

In their host species, rust fungi commonly elicit few of the wall-associated responses usually associated with plant cell wall penetration, despite the fact that they elicit a variety of such responses when they attempt to penetrate the walls of nonhost plants. In epidermal cells, these nonhost responses to rust fungi occur at an early stage of the penetration process, and penetration is usually arrested before the fungus comes into contact with the plant plasma membrane. These responses can be eliminated by interference with the plant actin cytoskeleton (Mellersh and Heath, 2001), suggesting that expression of wall responses requires an active "dialogue" between the plant cell wall and cytoplasm (Heath, 1997).

In mammalian cells, communication between the extracellular matrix and the cytoplasm is maintained via interactions between extracellular proteins containing arginine-glycine-aspartic acid (RGD) motifs and plasma membrane receptors called integrins (Giancotti and Ruoslahti, 1999). Exogenous application of peptides containing the RGD motif interferes with plasma membrane-cell wall adhesion in plant systems and also interferes with expression of wall-associated responses in fungal interactions with nonhost plants (Mellersh and Heath, 2001). Detailed cytological studies using sucrose-induced plasmolysis can reveal the degree of adhesion between plant cell walls and the plasma membrane as the latter pulls away during protoplast contraction. Under normal conditions, plants have numerous membrane-wall attachment points that are visualized by the presence of "Hechtian strands" following plasmolysis. Powdery mildew fungi appear to cause increases in the degree of membrane-wall adhesion, visualized by a change from a convex to a concave plasmolysis morphology, during infection of both host and nonhost plants (Lee-Stadelmann et al., 1984; Mellersh and Heath, 2001). Rust fungi, in contrast, cause no change in the strength in membrane wall adhesion, since they penetrate the walls of nonhost plants, and instead appear to locally decrease this adhesion under the penetration point in susceptible or resistant host plants (Mellersh and Heath, 2001). Thus it appears that rust fungi may locally disrupt adhesion as a means of interfering with the expression of potentially defensive wall responses in their host species. Determining the fungal molecule responsible for this decrease in adhesion will provide valuable insight into our understanding of host specificity for this important group of plant pathogens.

USE OF TRANSGENIC PLANTS TO STUDY EXPRESSION OF RESISTANCE AT THE CELLULAR LEVEL

Investigation of Cellular Resistance to Powdery Mildew Fungi Using Transient Expression Systems

Many of the plants for which we have the greatest understanding of cellular resistance are legumes and cereals. Unfortunately, these plants have proven difficult and time-consuming to genetically transform. As a result, many researchers have turned to the use of transient expression systems (Nelson and Bushnell, 1997; Schweizer et al., 1999). Biolistic transient expression systems are particularly well suited for situations in which fungi initially infect single epidermal cells, and thus have been used to clone elusive powdery mildew resistance genes from the complex *Mla* locus of barley (Halterman et al., 2001; Zhou et al., 2001), as well as to demonstrate the importance

(Halterman et al., 2001) or lack thereof (Zhou et al., 2001) of signaling components such as *Rar1* and associated *Rar*-interacting proteins (Azvedo et al., 2002) in the expression of resistance in individual cells.

Presence of the *mlo* allele in barley leads to broad-spectrum resistance to all *Erysiphe graminis* f. sp. *hordei* isolates tested and is manifested by early arrest of fungal development during the penetration stage (Schultze-Lefert and Vogel, 2000), in association with an enhanced localized plant response to the pathogen (Peterhansel et al., 1997; von Röpenack et al., 1998; Hückelhoven, Trujillo, and Kogel, 2000). Recently, single-cell transient expression of the wild-type *Mlo* gene in *mlo* barley reinstated susceptibility to the barley powdery mildew fungus, indicating that *mlo* functions in a cell-autonomous manner (Shirasu et al., 1999).

The wild-type *Mlo* gene encodes a 60 kDa protein (Büschges et al., 1997) that has features in common with the seven-transmembrane domain class of plasma-membrane receptors known as G-protein-coupled receptors (Devoto et al., 1999). However, transient expression of constitutively activated or inactivated variants of the barley G-protein α-subunit, or RNA-based silencing of the latter in single *mlo* or *Mlo* barley epidermal cells, had no effect on fungal penetration efficiency (Kim et al., 2002), suggesting *mlo* does not function through G-protein interactions. Instead, *Mlo* appears to exert its effect on penetration susceptibility through interactions with calmodulin, a calcium-binding protein. Although increases in cytosolic calcium are often associated with plant resistance (Xu and Heath, 1998), *Mlo* may actually condition cellular susceptibility through suppression of plant defenses in response to increased calcium levels.

CURRENT WORK IN OUR LABORATORY

Isolation of Fungal Effector Molecules

Cytological studies of rust fungal interactions with their host plants point to the cell wall as a primary site of determination of both susceptibility (Mould and Heath, 1999; Mellersh and Heath, 2001) and resistance (Xu and Heath, 1998; Mould and Heath, 1999). Rust fungi may produce proteinaceous molecules during the penetration stage aimed at interfering with the adhesion between the plant cell wall and plasma membrane, which is required for the expression of wall-associated responses. In order to identify the fungal molecule(s) responsible for this effect, we are using a *Saccharomyces cerevisiae*–based signal sequence trap technique in which 5' enriched cDNAs from developing rust fungal basidiospores are fused in frame with a truncated yeast invertase lacking an initiator methionine and signal sequence. cDNAs encoding the N-termini of proteins capable of directing this fusion protein to the secretory pathway are selected for because of their ability to allow an invertase-deficient strain of yeast to grow on sucrose medium and are likely to encode proteins relevant to the initial pathogenesis process, including the adhesion-reducing molecule(s). This screen may also be helpful in cloning peptide cell death elicitors produced by the cowpea rust fungus (D'Silva and Heath, 1997) that are produced following appressorium formation—the perception of which is believed to occur during penetration of the plant cell wall (Xu and Heath, 1998; Mould and Heath, 1999).

Early Changes in Transcription of Genes Related to Rust Resistance or Susceptibility

Recently, extraction and global amplification of mRNA from single infected resistant or susceptible host cells at the penetration stage of rust infection have revealed transcriptional changes for a number of genes (Mould and Heath, unpublished data). The absence of plant cell responses related to cell wall penetration at this stage of fungal infection should ensure that most of these transcriptional changes relate specifically to the induction of the hypersensitive response or to the establishment of cellular compatibility. We plan to manipulate expression of these genes in single infected and uninfected epidermal cells using a biolistic single-cell transient expression system in order to assess their functional relevance during host cell responses to rust fungi.

CONCLUDING REMARKS

Cell biological data support the plant cell wall as the primary line of defense against direct-penetrating fungi, especially in the case of nonhost plants. It is becoming increasingly clear that responses associated with the wall are subject to complex regulation by both plant and pathogen, and it remains a considerable challenge to determine which features are actually involved in stopping fungal growth. The cellular events surrounding resistance expressed following successful fungal penetration (i.e., the hypersensitive response) appear equally complex, and although in recent years have provided considerable insight into how these responses are initiated and executed in individual systems, truly comparable data of the responses among various plant-fungal systems remain scarce.

Recent advances in imaging techniques, such as computer enhanced video microscopy as well as confocal microscopy, have provided an opportunity to expand our knowledge of the cellular responses that occur in living cells (Heath, 2000a), including the activities of dynamic (and potentially short-lived) molecules such as ions and proteins. In addition, scientists can now manipulate these activities in single cells using transient expression systems and by fixing or displacing cellular components using laser-trapping techniques. Increasing biochemical, molecular, and proteomic data are available to serve as an important adjunct to cell biology. Together these techniques should serve to increase the importance of cell biology in our understanding of plant resistance to fungal parasites in the years ahead.

REFERENCES

Adam, L. and Somerville, S.C. (1996). Genetic characterization of five powdery mildew disease resistance loci in *Arabidopsis thaliana. The Plant Journal* 9: 341-356.

Aist, J.R. (1976). Papillae and related wound plugs of plant cells. *Annual Review of Phytopathology* 14: 145-163.

Aist, J.R. and Bushnell, W.R. (1991). Invasion of plants by powdery mildew fungi, and cellular mechanisms of resistance. In Cole, G.T. and Hoch, H.C. (Eds.), *The Fungal Spore and Disease Initiation in Plants and Animals* (pp. 321-345). New York and London: Plenum Press.

Aist, J.R. and Israel, H.W. (1977). Papilla formation: Timing and significance during penetration of barley coleoptiles by *Erysiphe graminis hordei. Phytopathology* 67: 455-461.

Azevedo, C., Sadanandom, A., Kitagawa, K., Freialdenhoven, A., Shirasu, K., and Schulze-Lefert, P. (2002). The RAR1 interactor SGT1, an essential component of *R* gene-triggered disease resistance. *Science* 295: 2073-2076.

Bolwell, G.P. (1999). Role of active oxygen species and NO in plant defence responses. *Current Opinion in Plant Biology* 2: 287-294.

Büschges, R., Hollricher, K., Panstruga, R., Simons, G., Wolter, M., Frijters, A., van Dailen, R., van der Lee, T., Diergaarde, P., Groenendijk, J., et al. (1997). The barley *Mlo* gene: A novel control element of plant pathogen resistance. *Cell* 88: 695-705.

Carver, T.L.W., Lyngkjaer, M.F., Neyron, L., and Strudwicke, C.C. (1999). Induction of cellular accessibility and inaccessibility and suppression and potentiation of cell death in oat attacked by *Blumeria graminis* f. sp. *avenae. Physiological and Molecular Plant Pathology* 55: 183-196.

Carver, T.L.W., Robbins, M.P., Zeyen, R.J., and Dearne, G.A. (1992). Effects of PAL-specific inhibition on suppression of activated defense and quantitative susceptibility of oats to *Erysiphe graminis. Physiological and Molecular Plant Pathology* 41: 149-163.

Carver, T.L.W., Zeyen, R.J., and Ahlstrand, G.G. (1987). The relationship between insoluble silicon and success or failure of attempted primary penetration by powdery mildew (*Erysiphe graminis*) germlings on barley. *Physiological and Molecular Plant Pathology* 31: 133-148.

Carver, T.L.W., Zhang, L., Zeyen, R.J., and Robbins, M.P. (1996). Phenolic biosynthesis inhibitors suppress adult plant resistance to *Erysiphe graminis* in oat at 20°C and 10°C. *Physiological and Molecular Plant Pathology* 49: 121-141.

Chen, C.Y. and Heath, M.C. (1991). Cytological studies of the hypersensitive death of cowpea epidermal cells induced by basidiospore-derived infection by the cowpea rust fungus. *Canadian Journal of Botany* 69: 1199-1206.

Chen, C.Y. and Heath, M.C. (1994). Features of the rapid cell death induced in cowpea by the monokaryon of the cowpea rust fungus or the monokaryon-derived cultivar-specific elicitor of necrosis. *Physiological and Molecular Plant Pathology* 44: 157-170.

Delledonne, M., Zeier, J., Marocco, A., and Lamb, C. (2001). Signal interactions between nitric oxide and reactive oxygen intermediates in the plant hypersensitive disease resistance response. *Proceedings of the National Academy of Sciences, USA* 98: 13454-13459.

Devoto, A., Piffanelli, P., Nilsson, I., Wallin, E., Panstruga, R., von Heijne, G., and Schulze-Lefert, P. (1999). Topology, subcellular localization, and sequence diversity of the Mlo family in plants. *The Journal of Biological Chemistry* 274: 34993-35004.

D'Silva, I. and Heath, M.C. (1997). Purification and characterization of two novel hypersensitive response-inducing specific elicitors produced by the cowpea rust fungus. *The Journal of Biological Chemistry* 272: 3924-3927.

Eulgem, T., Rushton, P.J., Schmelzer, E., Hahlbrock, K., and Somssich, I.E. (1999). Early nuclear events in plant defence signalling: Rapid gene activation by WRKY transcription factors. *The EMBO Journal* 18: 4689-4699.

Fernandez, M.R. and Heath, M C. (1989). Interactions of the nonhost French bean plant (*Phaseolus vulgaris*) with parasitic and saprophytic fungi: III. Cytologically-detectable responses. *Canadian Journal of Botany* 67: 676-686.

Freytag, S., Arabatzis, N., Hahlbrock, K., and Schmelzer, E. (1994). Reversible cytoplasmic rearrangements precede wall apposition, hypersensitive cell death and defense-related gene activation in potato/*Phytophthora infestans* interactions. *Planta* 194: 123-135.

Giancotti, F.G. and Ruoslahti, E. (1999). Integrin signaling. *Science* 285: 1028-1032.

Gregersen, P.L., Thordal-Christensen, H., Förster, H., and Collinge, D.B. (1997). Differential gene transcript accumulation in barley leaf epidermis and mesophyll in response to attack by *Blumeria graminis* f. sp. *hordei* (syn. *Erysiphe graminis* f. sp. *hordei*). *Physiological and Molecular Plant Pathology* 51: 85-97.

Gus-Mayer, S., Naton, B., Hahlbrock, K., and Schmelzer, E. (1998). Local mechanical stimulation induces components of the pathogen defense response in parsley. *Proceedings of the National Academy of Sciences, USA* 95: 8398-8403.

Halterman, D., Zhou, F., Wei, F., Wise, R.P., and Schulze-Lefert, P. (2001). The Mla6 coiled-coil, NBS-LRR protein confers *AvrMla6*-dependent resistance specificity to *Blumeria graminis* f. sp. *hordei* in barley and wheat. *The Plant Journal* 25: 335-348.

Hammerschmidt, R. (1999). Phytoalexins: What have we learned after 60 years? *Annual Review of Phytopathology* 37: 285-306.

Hazen, B.E. and Bushnell, W.R. (1983). Inhibition of the hypersensitive reaction in barley to powdery mildew by heat shock and cytochalasin B. *Physiological Plant Pathology* 23: 421-438.

Heath, M.C. (1983). Relationship between developmental stage of the bean rust fungus and increased susceptibility of surrounding bean tissue to the cowpea rust fungus. *Physiological Plant Pathology* 22: 45-50.

Heath, M.C. (1997). Signalling between pathogenic rust fungi and resistant or susceptible host plants. *Annals of Botany* 80: 713-720.

Heath, M.C. (1998). Involvement of reactive oxygen species in the response of resistant (hypersensitive) or susceptible cowpeas to the cowpea rust fungus. *New Phytologist* 138: 251-263.

Heath, M.C. (2000a). Advances in imaging the cell biology of plant-microbe interactions. *Annual Review of Phytopathology* 38: 443-459.

Heath, M.C. (2000b). Hypersensitive response-related death. *Plant Molecular Biology* 44: 321-334.

Heath, M.C., Nimchuck, Z.L., and Xu, H. (1997). Plant nuclear migrations as indicators of critical interactions between resistant or susceptible cowpea epidermal cells and invasion hyphae of the cowpea rust fungus. *New Phytologist* 135: 689-700.

Heath, M.C. and Škalamera, D. (1997). Cellular interactions between plants and biotrophic fungal parasites. *Advances in Botanical Research* 24: 195-225.

Heath M.C. and Stumpf, M.A. (1986). Ultrastructural observations of penetration sites of the cowpea rust fungus in untreated and silicon-depleted French bean cells. *Physiological and Molecular Plant Pathology* 29: 27-39.

Hückelhoven, R., Dechert, C., and Kogel, K. (2001). Non-host resistance of barley is associated with a hydrogen peroxide burst at sites of attempted penetration by wheat powdery mildew fungus. *Molecular Plant Pathology* 2: 199-205.

Hückelhoven, R., Dechert, C., Trujillo, M., and Kogel, K. (2001). Differential expression of putative cell death regulator genes in near-isogenic, resistant and susceptible barley lines during interaction with the powdery mildew fungus, *Plant Molecular Biology* 47: 739-748. .

Hückelhoven, R., Fodor, J., Trujillo, M., and Kogel, K. (2000). Barley *Mla* and *Rar* mutants compromised in the hypersensitive cell death response against *Blumeria graminis* f. sp. *hordei* are modified in their ability to accumulate reactive oxygen intermediates at sites of fungal invasion. *Planta* 212: 16-24.

Hückelhoven, R. and Kogel, K. (1998). Tissue-specific superoxide generation at interaction sites in resistant and susceptible near-isogenic barley lines attacked by the powdery mildew fungus (*Erysiphe graminis* f. sp. *hordei*). *Molecular Plant-Microbe Interactions* 11: 292-300.

Hückelhoven, R., Trujillo, M., and Kogel, K. (2000). Mutations in *Ror1* and *Ror2* genes cause modifications of hydrogen peroxide accumulation in *mlo*-barley under attack from the powdery mildew fungus. *Molecular Plant Pathology* 1: 287-292.

Kim, M.C., Panstruga, R., Elliott, C., Müller, J., Devoto, A., Yoon, H.W., Park, H.C., Cho, M.J., and Schultze-Lefert, P. (2002). Calmodulin interacts with MLO protein to regulate defence against mildew in barley. *Nature* 416: 447-450.

Kobayashi, Y., Kobayashi, I., and Kunoh, H. (1993). Recognition of a pathogen and a nonpathogen by barley coleoptile cells: II. Alteration of cytoplasmic strands in coleoptile cells caused by the pathogen *Erysiphe graminis*, and the nonpathogen, *E. pisi*, prior to their penetration. *Physiological and Molecular Plant Pathology* 43: 243-254.

Kobayashi, Y., Yamada, M., Kobayashi, I., and Kunoh, H. (1997). Actin microfilaments are required for the expression of nonhost resistance in higher plants. *Plant Cell Physiology* 38: 725-733.

Kunoh, H., Hayashimoto, A., Harui, M., and Ishizaki, H. (1985) Induced susceptibility and enhanced resistance at the cellular level in barley coleoptiles: I. The significance of timing of fungal invasion. *Physiological Plant Pathology* 27: 43-54.

Kunoh, H., Kuroda, K., Hayashimoto, A., and Ishizaki, H. (1986). Induced susceptibility and enhanced resistance at the cellular level in barley coleoptiles: II. Timing and localization of induced susceptibility in a single coleoptile cell and its transfer to an adjacent cell. *Canadian Journal of Botany* 64: 889-895.

Lee-Stadelmann, O.Y., Bushnell, W.R., and Stadelmann, E.J. (1984). Changes of plasmolysis form in epidermal cells of *Hordeum vulgare* infected by *Erysiphe graminis*: Evidence for increased membrane-wall adhesion. *Canadian Journal of Botany* 62: 1714-1723.

Lu, H. and Higgins, V.J. (1999). The effect of hydrogen peroxide on the viability of tomato cells and of the fungal pathogen *Cladosporium fulvum*. *Physiological and Molecular Plant Pathology* 54: 131-143.

Lyngkjaer, M.F., and Carver, T.L.W. (2000). Conditioning of cellular defence responses to powdery mildew in cereal leaves by prior attack. *Molecular Plant Pathology* 1: 41-49.

Lyngkjaer, M.F, Carver, T.L.W., and Zeyen, R.J. (2001). Virulent *Blumeria graminis* infection induces penetration susceptibility and suppresses race-specific hypersensitive resistance against avirulent attack in *Mla1*-barley. *Physiological and Molecular Plant Pathology* 59: 243-256.

Mansfield, J.W., Hargreaves, J.A., and Boyle, F.C. (1974). Phytoalexin production by live cells in broad bean leaves infected with *Botrytis cinerea*. *Nature* 252: 316-317.

McLusky, S.R., Bennett, M.H., Beale, M.H., Lewis, M.J., Gaskin, P., and Mansfield, J.W. (1999). Cell wall alterations and localized accumulation of feruloyl-3'-methoxytyramine in onion epidermis at sites of attempted penetration by *Botrytis allii* are associated with actin polarisation, peroxidase activity and suppression of flavonoid biosynthesis. *The Plant Journal* 17: 523-534.

Mellersh, D.G., Foulds, I.V., Higgins, V.J., and Heath, M.C. (2002). H_2O_2 plays different roles in determining penetration failure in three diverse plant-fungal interactions. *The Plant Journal* 29: 257-268.

Mellersh, D.G. and Heath, M.C. (2001). Plasma membrane-cell wall adhesion is required for expression of plant defense responses during fungal penetration, *The Plant Cell* 13: 413-424.

Mould, M.J.R. and Heath, M.C. (1999). Ultrastructural evidence of differential changes in transcription, translation, and cortical microtubules during *in planta* penetration of cells resistant or susceptible to rust infection. *Physiological and Molecular Plant Pathology* 55: 225-236.

Nelson, A.J. and Bushnell, W.R. (1997). Transient expression of anthocyanin genes in barley epidermal cells: Potential for use in evaluation of disease response genes. *Transgenic Research* 6: 233-244.

Nicholson, R.L. and Hammerschmidt, R. (1992). Phenolic compounds and their role in disease resistance. *Annual Review of Phytopathology* 30: 369-389.

Peng, M. and Kuc, J. (1992). Peroxidase-generated hydrogen peroxide as a source of antifungal activity in vitro and on tobacco leaf disks. *Phytopathology* 82: 696-699.

Perumalla, C.J. and Heath, M.C. (1989). Effect of callose inhibition on haustorium formation by the cowpea rust fungus in the non-host bean plant. *Physiological and Molecular Plant Pathology* 35: 375-382.

Perumalla, C.J. and Heath, M.C. (1991). The effect of inhibitors of various cellular processes on the wall modifications induced in bean leaves by the cowpea rust fungus. *Physiological and Molecular Plant Pathology* 38: 293-300.

Peterhansel, C., Freialdenhoven, A., Kurth, J., Kolsch, R. and Schulze-Lefert, P. (1997). Interaction analyses of genes required for resistance responses to powdery mildew in barley reveal distinct pathways leading to leaf cell death. *The Plant Cell* 9: 1397-1409.

Russo, V.M. and Bushnell, W.R. (1989). Responses of barley cells to puncture by microneedles and to attempted penetration by *Erysiphe graminis* f. sp. *hordei*. *Canadian Journal of Botany* 67: 2912-2921.

Schultze-Lefert, P. and Vogel, J. (2000). Closing the ranks to attack by powdery mildew. *Trends in Plant Science* 5: 343-348.

Schweizer, P., Pokorny, J., Abderhalden, O., and Dudler, R. (1999). A transient assay system for the functional assessment of defense-related genes in wheat. *Molecular Plant-Microbe Interactions* 12: 647-654.

Sherwood, R.T. and Vance, C.P. (1980). Resistance to fungal penetration in the Gramineae. *Phytopathology* 70: 273-279.

Shirasu, K., Nakajima, H., Rajasekhar, V.K., and Dixon, R.A. (1997). Salicylic acid potentiates an agonist-dependent gain control that amplifies pathogen signals in the activation of defense responses. *The Plant Cell* 9: 1-10.

Shirasu, K., Nielsen, K., Piffanelli, P., Oliver, R., and Schulze-Lefert, P. (1999). Cell-autonomous complementation of *mlo* resistance using a biolistic transient expression system. *The Plant Journal* 17: 293-299.

Škalamera, D. and Heath, M.C. (1998). Changes in the cytoskeleton accompanying infection-induced nuclear movements and the hypersensitive response in plant cells invaded by rust fungi. *The Plant Journal* 16: 191-200.

Škalamera, D., Jibodh, S., and Heath, M.C. (1997). Callose deposition during the interaction between cowpea *(Vigna unguiculata)* and the monokaryotic stage of the cowpea rust fungus *(Uromyces vignae)*. *New Phytologist* 136: 511-524.

Smart, M.G., Aist, J.R., and Israel, H.W. (1986). Structure and function of wall appositions: 2. Callose and the resistance of oversize papillae to penetration by *Erysiphe graminis* f. sp. *hordei*. *Canadian Journal of Botany* 64: 802-804.

Smart, M.G., Aist, J.R., and Israel, H.W. (1987). Some exploratory experiments on the permeability of papillae induced in barley coleoptiles by *Erysiphe graminis* f. sp. *hordei*. *Canadian Journal of Botany* 65: 745-749.

Snyder, B.A. and Nicholson, R.L. (1990). Synthesis of phytoalexins in sorghum as a site-specific response to fungal ingress. *Science* 248: 1637-1639.

Stakman, E.C. (1915). Relations between *Puccinia graminis* and plants highly resistant to its attack. *Journal of Agricultural Research* 4: 193-199.

Stumpf, M.A. and Heath, M.C. (1985). Cytological studies of the interactions between the cowpea rust fungus and silicon-depleted French bean plants. *Physiological Plant Pathology* 27: 369-385.

Thordal-Christensen, H., Zhang, Z., Wei, Y., and Collinge, D.B. (1997). Subcellular localization of H_2O_2 in plants. H_2O_2 accumulation in papillae and hypersensitive response during the barley-powdery mildew interaction. *The Plant Journal* 11: 1187-1194.

Tomiyama, K. (1960). Some factors affecting the death of hypersensitive potato plant cells infected by *Phytophthora infestans*. *Phytopathologische Zeitschrift* 39: 134-148.

Tomiyama, K. (1967). Further observation on the time requirement for hypersensitive cell death of potatoes infected by *Phytophthora infestans* and its relation to metabolic activity. *Phytopathologische Zeitschrift* 58: 367-378.

von Röpenack, E., Parr, A., and Schulze-Lefert, P. (1998) Structural analyses and dynamics of soluble and cell wall-bound phenolics in a broad spectrum resistance to the powdery mildew fungus in barley. *The Journal of Biological Chemistry* 273: 9013-9022.

Xiao, S., Ellwood, S., Calis, O., Patrick, E., Li, T., Coleman, M., and Turner, J.G. (2001). Broad-spectrum resistance in *Arabidopsis thaliana* mediated by *RPW8*. *Science* 291: 118-120.

Xu, H. and Heath, M.C. (1998). Role of calcium in signal transduction during the hypersensitive response caused by basidiospore-derived infection of the cowpea rust fungus. *The Plant Cell* 10: 585-597.

Yamaoka, N., Toyoda, K., Kobayashi, I., and Kunoh, H. (1994). Induced accessibility and enhanced inaccessibility at the cellular level in barley coleoptiles: XIII. Significance of haustorium formation by the pathogen *Erysiphe graminis* for induced accessibility to the non-pathogen *E. pisi* as assessed by nutritional manipulations. *Physiological and Molecular Plant Pathology* 44: 217-225.

Zhou, F., Kurth, J., Wei, F., Elliott, C., Valè, G., Yahiaoui, N., Keller, B., Somerville, S., Wise, R., and Schulze-Lefert, P. (2001). Cell-autonomous expression of barley *Mla1* confers race-specific resistance to the powdery mildew fungus via a *Rar1*-independent signaling pathway. *The Plant Cell* 13: 337-350.

Chapter 3

The Hypersensitive Response and Its Role in Disease Resistance

Hans-Peter Stuible
Erich Kombrink

INTRODUCTION

Plants have evolved a large variety of sophisticated and efficient defense mechanisms to prevent the colonization of their tissues by microbial pathogens and parasites. Preformed physical and chemical barriers constitute the first line of defense. Superimposed upon these is a series of inducible defense responses that are initiated after successful recognition of the invading pathogen. The induced defense responses can be assigned to three major categories, according to their distinct temporal and spatial expression patterns (Kombrink and Somssich, 1995). These categories are as follows:

1. Immediate, early defense responses are initiated in the directly invaded plant cell upon recognition of the pathogen with signals being transduced to neighboring cells, frequently leading to rapid death of few challenged host cells, the so-called hypersensitive response (HR).
2. Subsequent to recognition, local activation of genes occurs in close vicinity of infection sites, resulting in numerous biochemical and metabolic plant modifications that are well conserved among different plant-pathogen interactions. They include the de novo synthesis of proteins involved in the formation of antimicrobial compounds, so-called phytoalexins, structural proteins incorporated into the cell wall, and miscellaneous protective proteins.

3. Systemic activation of genes encoding pathogenesis-related (PR) proteins, including chitinases and β-1,3-glucanases (see Chapter 5), which are directly or indirectly inhibitory toward pathogens, occurs temporarily delayed in plant tissues at a distance from the initial infection site, resulting in the establishment of an immunity to secondary infections termed systemic acquired resistance (SAR) (Sticher et al., 1997).

A common response associated with plant disease resistance is the HR, which occurs nearly ubiquitously in incompatible plant-pathogen interactions. It is manifested as rapid collapse and death of host cells and can be recognized within a few hours after contact between plant and pathogen occurs. Recent genetic, biochemical, and morphological evidence clearly suggests that HR cell death in plants is controlled by endogenous genetic mechanisms and hence is a kind of programmed cell death (PCD), similar to apoptosis of mammalian cells (Morel and Dangl, 1997; Heath, 1998; Danon et al., 2000). Nevertheless, it remains to be established to what extent cell death during HR is similar to or different from cell death that also occurs during other physiological and developmental stages of the plant's life cycle, for example, during differentiation of xylem tracheary elements, sexual reproduction, or senescence of specific plant organs such as petals or leaves (Pennell and Lamb, 1997). Another challenging question is whether the HR cell death invoked early in the resistance response is fundamentally different from HR-like cell death that occurs at relatively late time points during development of disease symptoms in compatible plant-pathogen interactions. It is a popular and common view that HR cell death has an important defense function and actually causes disease resistance by depriving the invading pathogen of nutrients and/or by releasing antimicrobial compounds from the dying cells. However, in several experimental systems, a separation of disease resistance from cell death formation has been demonstrated, indicating that successful restriction of pathogen spread and HR cell death are not as tightly linked as previously thought (Richael and Gilchrist, 1999).

In this chapter, we will summarize the common features of HR cell death as it occurs in incompatible interactions between plant and pathogen and review the recent evidence providing some insight into the regulatory mechanisms controlling its execution and its function in disease resistance. It is not our intention to provide a comprehensive review of the extensive literature on PCD or apoptosis, which is presently an intensely studied area. For various aspects, the reader is referred to excellent recent review articles (Hammond-Kosack and Jones, 1996; Morel and Dangl, 1997; Heath, 2000; Shirasu and Schulze-Lefert, 2000; Lam et al., 2001).

FEATURES OF THE HYPERSENSITIVE RESPONSE

The term *hypersensitive* was introduced by Stakman as early as 1915 (Stakman, 1915) to describe the rapid and localized cell death associated with cereal resistance to the rust fungus *Puccinia graminis*. Subsequently, the HR was recognized as a general defense reaction in numerous plant-pathogen interactions, irrespective of the inducing pathogen. It occurs in resistant plants in response to pathogenic viruses, bacteria, fungi, or nematodes and is associated with many morphological and biochemical changes. However, which of these changes are causally related to HR induction and which are the consequences of cells dying from infection is not totally clear. In view of apparent similarities to PCD in animal systems, considerable research effort in recent years has been devoted to determining to what extent hallmarks of animal apoptosis or PCD can be identified in plants undergoing HR cell death (Heath, 1998; Danon et al., 2000).

Apoptosis in animal cells is usually associated with a number of morphological features, such as chromatin condensation, nuclear blebbing, and cell fragmentation into apoptotic bodies, and at the very end it results in the ordered disintegration of the cell (Heath, 1998; Danon et al., 2000). These hallmarks are accompanied by biochemical changes, such as specific DNA cleavage to result in distinct ladders of fragments, activation of specific proteases (caspases) that activate downstream reactions, degradation of poly (ADP-ribose) polymerase that is believed to be involved in DNA repair, and release of cytochrome c from mitochondria. Many of these morphological and biochemical markers have been identified in various plant systems that exhibit developmentally controlled cell death or HR cell death in response to infection or treatment with toxins (Danon et al., 2000). In particular, DNA ladders and caspaselike activities have been described during HR formation following bacterial and fungal infection, whereas other markers, such as cytochrome c release or cleavage of poly (ADP-ribose) polymerase, have been detected in plant cells dying from heat or oxidative stress (Heath, 1998; Danon et al., 2000; Lam and del Pozo, 2000). In general, plant cells dying during the HR show several but not all hallmarks characteristic of animal PCD.

In animals, recent progress in research into PCD has resulted in the identification of proteins and protein domains involved in this process (Aravind et al., 1999). Most important, cell death is controlled by a variety of components, including receptors, adaptors, and downstream effectors, that interact via distinct protein interaction domains in different combinations. On the one hand, striking structural similarities between mediators or regulators of PCD in animals (e.g., Ced-4 and Apaf-1) and proteins encoded by plant resistance genes have been uncovered (Aravind et al., 1999; Dangl and Jones, 2001). On the other hand, key regulators such as members of the Bcl-2 family of proteins that either activate (e.g., Bax) or inhibit (e.g., Bcl-2) PCD in animals have until now not been identified in plants, nor have corresponding open reading frames been found in the complete genome of *Arabidopsis thaliana* (Danon et al., 2000; Lam and del Pozo, 2000; Lam et al., 2001). From these results, a picture emerges that cell death pathways in animals and plants share several common features, while at the same time apparent differences cannot be neglected. Where appropriate, various aspects related to the functional similarity or dissimilarity between components of animal PCD and plant HR will be covered in several subtopics as follows.

PROTEINS INVOLVED IN PATHOGEN RECOGNITION AND INITIATION OF HR CELL DEATH

Structure and Function of Plant Resistance Proteins

Disease resistance in plants is often mediated by specific interactions between plant resistance *(R)* genes and corresponding avirulence *(avr)* genes of the pathogen (Dangl and Jones, 2001). When corresponding *R* and *avr* genes are present in both host and pathogen, the result is disease resistance; if either gene is inactive or absent, the result is disease. The simplest model to explain this gene-for-gene interaction suggests a direct binding event between the plant R protein (receptor) and the pathogen-derived Avr product (ligand), which activates a signal-transduction cascade leading to HR. Alternatively, R proteins might detect modifications or conformational changes of primary targets of Avr proteins, rather that the Avr proteins themselves, a concept known as the "guard hypothesis" (Dangl and Jones, 2001). The term *"avr* gene" is somewhat misleading, as it suggests that an avirulence function is encoded, which does not seem beneficial for the pathogen. The evolutionary conservation of *avr* genes in pathogen genomes as well as recent experimental data suggest that proteins encoded by *avr* genes play an important role in the colonization of nonresistant host plants and hence are essential virulence factors of successful pathogens (Kjemtrup et al., 2000). Correspondingly, resistant plants have evolved efficient and fast mechanisms to recognize these virulence factors (*avr* gene products) and by mounting appropriate defense mechanisms are able to restrict pathogen proliferation and avoid disease.

In past years, many *R* genes have been isolated from model and crop plants (Dangl and Jones, 2001). Depending on the presence of typical structural motifs, such as a nucleotide-binding domains (NBs), leucine-rich repeats (LRRs), transmembrane domains (TMs), and serine/threonine protein kinase domains (PKs), the *R* genes presently known encode five classes of proteins (Dangl and Jones, 2001) (see Chapter 4).

The rice *Xa21* gene confers resistance to *Xanthomonas oryzae* carrying the avirulence gene *avrXa2* and represents a well-characterized member of the LRR-TM-PK-type of R protein. Interestingly, several developmental genes from *Arabidopsis thaliana,* such as *ERECTA* and *CLAVATA1,* and the gene encoding the brassinosteroid receptor BRI1, exhibit a domain organization comparable with *Xa21* (Jones, 2001). Chimeric gene fusions consisting of the extracellular LRR plus the TM domain of *BRI1* and the intracellular kinase domain of *Xa21* activated defense responses in cultured rice (*Oryza sativa* L.) cells after treatment with brassinosteroids (He et al., 2000). From these results, it has been concluded that different LRR domains recognize their distinct ligands and that the protein kinase domain specifies the utilized signal-transduction pathway (He et al., 2000).

The tomato (*Lycopersicon esculentum* Mill.) *Cf-X* genes represent another class of *R* genes that encode transmembrane proteins with extracellular LRRs and short cytoplasmic domains (Dangl and Jones, 2001). The *Cf-4* and *Cf-9* genes share high sequence identity (90 percent at the amino acid level) and specify resistance to races of the fungus *Cladosporium fulvum* carrying the *avr4* and *avr9* genes, respectively (Jones, 2001). An extensive series of domain swapping and gene shuffling experiments between *Cf-4* and *Cf-9* uncovered that recognition specificity resides mainly in the LRRs and may involve many *(Cf-9)* or only distinct repeats *(Cf-4)* (Jones, 2001).

The tomato *Pto* gene, conferring resistance to *Pseudomonas syringae* carrying *avrPto,* is unique in that it encodes a Ser/Thr kinase and consists of a PK domain only, whereas putative Avr recognition domains (LRRs) are absent. This unusual structural feature indicates that Pto may have the capacity to activate a signal transduction cascade by phosphorylation, whereas direct signal perception via LRR-mediated protein-protein interaction is not possible (Dangl and Jones, 2001). Nonetheless, Pto represents one of the rare R proteins, for which a direct interaction with its cognate avirulence gene product has been demonstrated (Tang et al., 1996). Based on the domain structure of Pto and the fact that it requires the NB-LRR protein Prf for its function, it has been postulated that Pto is a more general component of the plant defense system which is actively targeted by AvrPto to support growth of the invading pathogen. In this model, Prf functions as a guard that detects modifications or conformational changes of Pto by AvrPto, which then initiates the HR and other defense responses (Dangl and Jones, 2001).

The largest class of *R* genes encodes NB-LRR proteins, which are localized in the cytosol or may be associated with the plasma membrane (Dangl and Jones, 2001). The NB domains of these plant *R* gene products share significant similarities with domains localized in nematode Ced-4 and mammalian Apaf-1 proteins, which both function as cell death effectors and are involved in the activation of apoptotic proteases such as caspase-9 (Aravind et al., 1999; Dangl and Jones, 2001). Because the Avr/R interaction commonly induces

HR, this observation strongly suggests that functional similarities exist between PCD in animals and the HR cell death in plants.

NB-LRR proteins can be subdivided into two classes according to their structural features. The N-terminal region of one subclass exhibits homology with the intracellular signaling domains of the *Drosophila* Toll and mammalian interleukin-1 receptors (TIR). R proteins of this subclass are therefore designated TIR-NB-LRR proteins. In animal systems, Toll-like receptors are commonly responsible for recognition of so-called PAMPs (pathogen-associated molecular patterns), such as lipoproteins, peptidoglycans, lipopolysaccharides, and flagellin (Suzuki et al., 2002). The second subclass of NB-LRR proteins, instead of a TIR domain, carries a putative coiled-coil (CC) domain at the N-terminus. Correspondingly, these R proteins are designated CC-NB-LRR proteins (also referred to as leucine-zipper-class, LZ-NB-LRR proteins). CC or LZ domains are thought to facilitate protein dimerization. However, dimers of R proteins have not yet been reported.

An astonishing feature of NB-LRR *R* gene organization is the existence of extensive allelic series of resistance specificities at several individual loci scattered throughout the genomes analyzed so far (Dangl and Jones, 2001; Jones, 2001). One example for such exceptional diversification of *R* genes is the barley (*Hordeum vulgare* L.) *Mla* locus, encoding more than 30 different resistance specificities (*Mla-1* to *Mla-32*) against different isolates of barley powdery mildew, *Blumeria graminis* f. sp. *hordei* (Schulze-Lefert and Vogel, 2000). Within a genomic interval of 240 kb, at least 11 genes encoding NB-LRR proteins have been identified that fall into three distinct classes, one of which represents CC-NB-LRR proteins, and the other two truncated variants thereof (Halterman et al., 2001). The complete genome sequence of *Arabidopsis* has uncovered an even more complex structure of *R* genes in this plant, with more than 100 *R* loci scattered over all chromosomes, and more than 150 sequence homologues of the NB-LRR-type of *R* genes (Dangl and Jones, 2001). Although the evolutionary mechanisms of *R* gene diversification are still a point of discussion, the opportunity to compare an allelic series of R proteins represents an important source of information for unraveling the functions of their distinct domains. This might then allow the design of new *R* gene specificities by in vitro recombination.

Successful domain-swap experiments between *R* alleles have indicated that specificity largely resides in the C-terminal LRR repeats, as demonstrated for different types of resistance genes, including flax *L,* rice *Xa21,* and tomato *Cf-9.* However, a contribution of the N-terminal and C-terminal regions cannot be dismissed (Jones, 2001).

Additional Components Required for R Protein Function

Mutational analyses in several plants have defined loci that are required for *R* gene function. In the model plant *Arabidopsis thaliana,* several signal-transduction components were identified that connect *R*-gene-specific pathogen recognition with the initiation of HR cell death. One important result from these studies is that R proteins of the CC-NB-LRR and TIR-NB-LRR type utilize different disease-resistance signaling pathways (Feys and Parker, 2000; Shirasu and Schulze-Lefert, 2000; Dangl and Jones, 2001). The *eds1* mutant was identified in a screen for loss of race-specific resistance to the oomycete *Peronospora parasitica,* specified by the *RPP5* gene (TIR-NB-LRR-type) in the *Arabidopsis* accession Landsberg *erecta.* Further studies revealed a more general dependence of TIR-NB-LRR, but not CC-NB-LRR, R proteins on a functional EDS1 protein (Feys et al., 2001). Since EDS1 contains a domain with significant similarity to eukaryotic lipases, the protein may function by hydrolyzing or binding a lipidlike molecule. However, neither enzyme activity nor identity of a potential substrate has yet been established (Falk et al., 1999).

PAD4 is another lipaselike *Arabidopsis* protein required for the proper function of TIR-NB-LRR proteins (Feys et al., 2001). The *pad4* mutant was originally identified in a screen for plants with enhanced disease susceptibility to a virulent strain of the bacterium *Pseudomonas syringae* and was subsequently characterized as a component necessary for phytoalexin (i.e., camalexin) and salicylic acid (SA) accumulation (Glazebrook, 2001). Recently, it was demonstrated that EDS1 can dimerize or form heterodimers with PAD4 and that both proteins are required for accumulation of the signaling molecule salicylic acid (Feys et al., 2001). Nonetheless, both proteins apparently fulfill distinct functions in defense-signaling pathways. EDS1 is an essential component of HR formation and is associated with early plant defense responses, such as the oxidative burst. Therefore, in *eds1* mutants, HR cell death is suppressed and pathogen growth is not restricted. In contrast, PAD4 appears to strengthen or multiply an activity downstream of HR initiation. Thus, in *pad4* mutants (EDS1 wild-type), HR formation does occur but is insufficient to fully restrict pathogen growth, resulting in a typical phenotype, the formation of trailing necrotic lesions (Rustérucci et al., 2001).

NDR1, in contrast to EDS1 and PAD4, is required for the function of a subset of R proteins of the CC-NB-LRR type, such as RPM1 and RPT2. Mutations in NDR1 do not compromise TIR-NB-LRR protein-dependent resistant reactions, indicating that EDS1/PAD4 and NDR1 differentiate *R*-gene-mediated events conditioned by particular R protein structural types (Feys and Parker, 2000). NDR1, a basic protein with two putative membrane attachment domains, was identified by screening for loss of RPM1-mediated resistance to the bacterial pathogen *Pseudomonas syringae* carrying *avrB* (Century et al., 1995, 1997). Interestingly, both resistance and HR formation mediated by the *RPT2* gene are strictly dependent on a functional NDR1 protein, while the *RPM1* gene can trigger an HR-like reaction even in *ndr1* mutants, although growth of bacteria carrying *avrRPM1* or *avrB* is not restricted (Century et al., 1995). These observations indicate that certain R proteins might interact with different downstream signaling components or that HR formation and resistance could be separated under certain circumstances.

In barley, many of the *Mla*-specified resistances against *Blumeria graminis* f. sp. *hordei* (e.g., *Mla6*, *Mla9*, *Mla12*) require for their function at least two additional genes, called *Rar1* and *Rar2* (Schulze-Lefert and Vogel, 2000). The *rar1-1* and *rar1-2* mutations were originally isolated as suppressors of *Mla12* (Jørgensen, 1996). However, the most remarkable recent finding is that structurally highly conserved *Mla* alleles differ in their requirement for *Rar1,* which encodes a novel intracellular protein with two zinc-binding motifs, designated CHORD (cysteine- and histidine-rich domain), that are also found in proteins from other higher organisms (Shirasu et al., 1999). For example, *Mla1,* in contrast to *Mla6,* confers *Rar1*-independent resistance, although both *R* genes encode proteins of the CC-NB-LRR type with 91 percent sequence identity (Halterman et al., 2001; Zhou et al., 2001). This points to the existence of more than one race-specific signaling pathway induced by different R proteins of the same structural type and raises the question of how such a function is exerted by structurally highly homologous NB-LRR proteins.

During *Mla12*-specified resistance of barley to *Blumeria graminis* f. sp. *hordei*, HR cell death usually occurs at 24 h postinoculation in attacked epidermal cells and is preceded by an H_2O_2 burst (Hückelhoven et al., 1999; Shirasu et al., 1999). Mutants in *Rar1* and *Rar2* are compromised in epidermal H_2O_2 accumulation and fail to execute HR cell death and pathogen resistance (Shirasu et al., 1999). This places *Rar1* upstream of an H_2O_2 burst that might drive the attacked cells into programmed suicide.

SMALL MOLECULES INVOLVED IN SIGNALING AND EXECUTION OF HR CELL DEATH

One of the most rapid plant responses engaged following pathogen recognition is the oxidative burst, which constitutes the production of reactive oxygen intermediates (ROI), primarily superoxide (O_2^-) and hydrogen peroxide (H_2O_2), at the site of attempted invasion (Lamb and Dixon, 1997; Wojtaszek, 1997; Grant and Loake, 2000). It has been suggested that the oxidative burst and cognate redox signaling may play a central role in integration and coordination of the multitude of plant defense responses (Lamb and Dixon, 1997).

Superoxide anion generation in relation to HR was first reported for potato tuber slices inoculated with an avirulent race of *Phytophthora infestans* (Doke, 1983). Subsequently, the oxidative burst has been identified in numerous plant-pathogen interactions involving different kinds of pathogens. The origin of ROI generated during the oxidative burst is not unequivocally established, but candidate reactions are the action of a plasma membrane-located NADPH-dependent oxidase complex and cell wall peroxidases (Wojtaszek, 1997; Grant and Loake, 2000). The cytotoxicity and reactive nature of O_2^- requires its cellular concentration to be carefully controlled, which can be achieved by induction of antioxidant enzymes, such as glutathione *S*-transferase, glutathione peroxidase, or ascorbate peroxidase (Wojtaszek, 1997; Smirnoff, 2000).

ROI accumulation is a rapid event that precedes HR cell death in many plant-pathogen interactions showing *R*-gene-triggered resistance (Kombrink and Somssich, 1995; Wojtaszek, 1997). Rapid and biphasic ROI accumulation has been observed in several cultured plant cell systems in response to bacterial or fungal elicitors, i.e., *avr* gene products (Levine et al., 1994; Jabs et al., 1997; Wojtaszek, 1997). While the first peak was considered nonspecific, the second sustained ROI burst was dependent on the pathogen race and only occurred with avirulent bacteria. Collectively, these data suggest a dual function for ROI in disease resistance: (1) direct participation in the development of host cell death during HR as well as direct inhibition of the pathogen, and (2) a role as a diffusable signal for induction of cellular protectants and defense responses in neighboring cells. Thus the strict spatial limitation of HR cell death may be the result of a dose-dependent action of ROI (Lamb and Dixon, 1997).

In animal cells, nitric oxide (NO) is known to act as second messenger in concert with ROI in processes such as the innate immune response, inflammation, and PCD. Recently it was shown that NO might also play an important role in the regulation of defense responses in plants. Infection of resistant, but not susceptible, tobacco plants with tobacco mosaic virus (TMV) resulted in enhanced NO synthase activity (Durner et al., 1998). Furthermore, external application of NO induced salicylic acid accumulation and PR gene expression, and NO inhibitors blocked both effects. This suggests that several critical players of animal NO signaling, such as cyclic GMP or cyclic AMP-ribose, are also operative in plants (Durner et al., 1998; Klessig et al., 2000). In cultured soybean (*Glycine max* L.) cells, it was demonstrated that the efficient induction of HR cell death required a balance between ROI and NO production, whereas unregulated NO production was not sufficient to induce HR cell death (Delledonne et al., 2001).

Other rapid changes observed following pathogen recognition are selective ion fluxes across the plasma membrane (Kombrink and Somssich, 1995; Nürnberger and Scheel, 2001). Although rapid responses have been extensively studied in appropriate model systems, such as cultured cells stimulated with defined elicitors, some debate concerns the precise temporal order in which they occur. In cultured parsley cells, it has been established that ion fluxes (H^+, K^+, Cl^-,

Ca^{2+}) precede the oxidative burst (Jabs et al., 1997), whereas in cultured soybean (*Glycine max* L.) cells, the oxidative burst apparently precedes and stimulates a rapid influx of Ca^{2+}, which then leads to HR cell death (Levine et al., 1996; Morel and Dangl, 1997). The specific requirement of calcium signaling in HR cell death had been suggested from studies using ionophores and calcium-channel blockers (Jabs et al., 1997; Nürnberger and Scheel, 2001). Recent work suggests that this dependence on Ca^{2+} may involve specific isoforms of calmodulin, since HR-like cell death, PR protein gene expression, and broad-spectrum disease resistance were induced by transgenic expression of soybean calmodulin-encoding genes in tobacco (Heo et al., 1999). The molecular mechanisms of these calmodulin-induced responses are not yet known; however, it was recently demonstrated that calmodulin directly interacts with the barley Mlo protein and regulates defense against powdery mildew (Kim et al., 2002).

The function of salicylic acid as a crucial signal molecule that is involved in systemic acquired resistance and PR gene expression has been known for many years (Sticher et al., 1997). More recently, SA has also emerged as a positive feedback regulator of cell death during the HR (Feys and Parker, 2000). This was first suggested from studies with lesion mimic mutants of *Arabidopsis* (Weymann et al., 1995). The spontaneous cell death formation in the *lsd6* and *lsd7* mutants was suppressed by transgenic expression of the *NahG* gene, encoding a bacterial SA hydroxylase that degrades SA to catechol. Additional lesion-mimic mutants, *ssi1* and *acd6,* that were recently isolated also show a SA-dependent HR phenotype (Rate et al., 1999; Shah et al., 1999). Furthermore, SA depletion in tobacco by *NahG* expression delayed HR cell death after infection with avirulent bacteria, and this delay was correlated with a reduced and delayed oxidative burst (Draper, 1997). Taken together, these data suggest that SA, in addition to SAR signaling, also has a role in early and local defense regulation by amplifying and sustaining the oxidative burst. In fact, SA might act in concert with ROI to define the threshold required for initiation of HR cell death.

NEGATIVE REGULATORY COMPONENTS

In order to survive pathogen attack, it is essential for the plant to establish fast and efficient defense responses, including HR cell death, but it is equally important to minimize tissue damage by limiting the HR to the few directly affected cells. Thus the sensitive mechanisms triggering HR cell death after infection have to be balanced by mechanisms that prevent HR in the absence of pathogens. Based on these considerations, it seems reasonable to assume that some components of defense-signaling pathways function as negative regulators of cell death, and likewise, that certain positive regulatory components are degraded after initiation of HR cell death in order to prevent deleterious spread of HR lesions beyond the infected area. Both assumptions are supported by recent evidence.

In *Arabidopsis thaliana,* the *R* gene product RPM1, conferring resistance to *Pseudomonas syringae* strains expressing either *avrRpm1* or *avrB,* is rapidly degraded during incompatible interactions, concomitant with the onset of the HR. This observation points to the existence of a negative feedback loop controlling the extent of cell death at the site of infection (Boyes et al., 1998).

A large number of mutants have been isolated from a variety of plant species, including *Arabidopsis,* maize, barley, and tomato, that show a spontaneous cell death phenotype even in the absence of a pathogen. Such mutants are commonly classified as lesion mimics, and in fact, some of them provide evidence for the existence of cell-death protection systems in plants that actively suppress the HR in uninfected tissue. Lesion mimics may or may not exhibit enhanced pathogen resistance, depending on the type of lesion and/or the lifestyle of the pathogen (Shirasu and Schulze-Lefert, 2000). Likewise, development of cell death is often associated with the induction of defense-related markers, such as callose deposition or PR gene expression.

In the *Arabidopsis lsd1* mutant, elevated resistance against normally virulent bacterial and fungal pathogens is associated with a spreading lesion phenotype; i.e., once initiated, lesions spread and finally consume the entire leaf (Dietrich et al., 1994, 1997). This type of mutant obviously fails to limit the extent of cell death formation, not only in response to infection but also upon exposure to other stimuli, such as extracellular application of superoxide or salicylic acid, which both initiate the so-called runaway cell death phenotype. The *LSD1* gene encodes a protein with three zinc finger domains, which may confer binding capacity for other proteins or nucleic acids, and therefore it was proposed that the LSD1 protein functions as a transcriptional activator (Dietrich et al., 1997). Since LSD1 was recently shown to be part of a signaling pathway leading to the induction of copper zinc superoxide dismutase, it has been suggested that the spreading lesion phenotype of *lsd1* results from an insufficient detoxification of accumulating superoxide and other ROI, which then trigger a cell death cascade even in unaffected tissue (Kliebenstein et al., 1999).

Results accumulating over the past few years have clearly shown that various types of protein kinases participate in the activation of plant defense responses, including HR cell death (Romeis, 2001). However, two recent examples indicate that MAP kinase (mitogen-activated protein kinase) cascade components may also function as negative regulators of defense responses. First, the *Arabidopsis mpk4* mutant, containing the transposon-inactivated MAP kinase 4, exhibited an increased resistance to virulent pathogens and constitutive systemic-acquired resistance, accompanied with elevated levels of salicylic acid and constitutive PR gene expression (Petersen et al., 2000). Thus the potential MAP kinase cascade utilizing MPK4 apparently has a negative regulatory role in the plant defense response. Second, the *Arabidopsis EDR1* gene was found to encode a putative MAPKKK (mitogen-activated protein kinase kinase kinase) with significant sequence similarity to CTR1, a negative regulator of ethylene responses (Frye et al., 2001). The recessive *edr1* mutant showed enhanced resistance against the bacterium *Pseudomonas syringae* and the powdery mildew fungus *Erysiphe cichoracearum*. Significantly, the *edr1* mutant does not constitutively express defense responses, unlike many other disease-resistant mutants of *Arabidopsis,* but after infection, callose deposition, PR gene expression, and HR cell death are induced much faster than in wild-type plants. Taken together, these results suggest that EDR1 functions at the top of a MAP kinase cascade that negatively regulates defense responses (Frye et al., 2001).

In barley, *mlo* alleles confer broad-spectrum disease resistance to nearly all races of the powdery mildew fungus *Blumeria graminis* f. sp. *hordei* (Schulze-Lefert and Vogel, 2000). Similar to *lsd1* in *Arabidopsis, mlo* is a recessive mutation in barley, and isolation of the *Mlo* gene confirmed that the resistance phenotype is indeed caused by loss-of-function mutations in the wild-type gene (Büschges et al., 1997). The 60 kDa Mlo protein is anchored in the plasma membrane by seven transmembrane domains, a topology and subcellular localization reminiscent of G-protein-coupled receptors (Devoto et al., 1999). However, such a function was dismissed by recent experimental evidence (Kim et al., 2002). Instead, it was demonstrated that Mlo interacts with calmodulin and that the loss of calmodulin binding partially impairs Mlo wild-type activity of modulating defense responses against pathogens (Kim et al., 2002). Mutations in each of two additional genes, *Ror1* and *Ror2,* compromise the resistance of *mlo* plants, suggesting that wild-type *Mlo* functions as a negative regulator of resistance responses (requiring *Ror1* and *Ror2*) and that the lack of *Mlo* primes defense responses in *mlo* plants (Schulze-Lefert and Vogel, 2000). This is further supported by the observation that *mlo* mutants form spontaneous lesions in the absence of a pathogen. The *mlo* lesion-mimic phenotype may further indicate the involvement of a threshold control for initiation of the HR pathway, which may be the final step of increasingly severe cellular resistance reactions, as previously proposed for the *lsd1* lesion mimic (Dietrich et al., 1994; Morel and Dangl, 1997).

IS PROTEIN DEGRADATION IMPORTANT FOR EXECUTION OF HR CELL DEATH?

As outlined earlier, increasing evidence suggests that removal of negative regulatory components may play an important role in the execution of plant HR cell death. In the case of proteinaceous factors, this could be achieved by proteases with high substrate specificity or alternatively by specific targeting of such proteins to the ubiquitin-proteasome degradation pathway.

It is well established that activation of regulatory protease cascades is required for the initiation of PCD in animal cells (Grütter, 2000). The most prominent family of proteases involved in this process carries a cysteine in their catalytic site and cleaves specifically after an aspartate residue; therefore, these proteases are called caspases (cysteine dependent aspartate-specific proteases). Caspases are (auto)activated from an inactive zymogen after binding of an extracellular signal molecule to its cognate receptor or as a consequence of intracellular stresses. After activation, the initiator caspases process additional downstream effector procaspases, which finally lead to the execution of cell death by removal of suppressors of apoptosis (Aravind et al., 1999; Grütter, 2000). In plants, proteases with significant sequence similarity to caspases have not yet been identified, nor are such proteins encoded in the complete *Arabidopsis* genome (Estelle, 2001). However, by using specific peptide substrates and selective protease inhibitors, caspaselike activities have recently been identified in plants. For example, caspaselike activities have recently been detected in tobacco tissue undergoing a virus-induced HR, as well as in the cytosol of embryonic barley cells and in the giant cells of the algae *Chara corallina* (Korthout et al., 2000; Lam and del Pozo, 2000). Another feature of PCD in animal cells is the specific cleavage of poly (ADP-ribose) polymerase (PARP) by caspase-3. Corresponding with the activation of caspase-3-like activity, PARP degradation has been observed in cultured tobacco cells upon heat treatment (Estelle, 2001).

Although data suggesting a participation of caspaselike proteases in regulation of cell death in plants are still indirect, increasing evidence suggests that regulatory proteins are specifically targeted to the ubiquitin/proteasome degradation pathway. In this pathway, ubiquitin becomes covalently attached to the protein destined for degradation by an ATP-dependent reaction cascade comprising three enzymes or enzyme complexes (Callis and Vierstra, 2000). E1 (ubiquitin-activating enzyme) catalyzes the formation of an activated ubiquitin that is subsequently transferred to the cysteinyl sulfhydryl group of a second enzyme called E2 (ubiquitin-conjugating enzyme). E2s represent a large family of proteins with at least 36 isoforms in *Arabidopsis* (Callis and Vierstra, 2000). The transfer of ubiquitin to its target protein requires a third specificity-conferring protein or protein complex called E3 (ubiquitin ligase). The highly diverse E3s have been grouped into four classes, including the SKP1-Cullin-F-box (SCF) E3 ligase subtype, which is named after the three yeast protein subunits: suppressor of kinetochore protein 1 (Skp1p), cell-division cycle 53 (Cullin), and F-box proteins (Callis and Vierstra, 2000). Following monoubiquitination, subsequent attachment of additional ubiquitin units to the primary ubiquitin residue leads to the formation of polyubiquitinated proteins, which are then targeted for degradation by the *26S* proteasome.

A direct link between resistance and protein ubiquitination has recently been discovered in barley and *Arabidopsis*. As outlined earlier, race-specific resistance of barley to powdery mildew that is mediated by some *Mla* alleles, such as *Mla6* and *Mla12,* requires the additional components *Rar1* and *Rar2.* To elucidate the molecular function of RAR1 in plants, the *Arabidopsis* homologue AtRAR1 was used as bait in a yeast two-hybrid screen to search for interacting proteins. Thereby, two interacting proteins were identified, AtSGT1a and AtSGT1b, which shared extensive sequence similarity to each other and to the yeast protein SGT1 (Azevedo et al., 2002). In yeast, SGT1 is an essential regulator of the cell cycle. Its function could be complemented by both AtSGT1a and AtSGT1b, suggesting that these proteins are functional orthologs of yeast SGT1. In addition, yeast SGT1 is associated with the kinetochore complex and the SCF-type (Skp1-Cullin-F-box) E3 ubiquitin ligase (Kitagawa et al., 1999). In barley, immunoprecipitation experiments demonstrated that SGT1 interacts not only with either RAR1 and SCF subunits, but also with two subunits of the COP9 signalosome, which are closely related to the lid complex of the *26S* proteasome (Karniol and Chamovitz, 2000; Azevedo et al., 2002). Thus one possible role of plant SGT1 could be to target resistance-regulating proteins for polyubiquitination and subsequent degradation by the *26S* proteasome.

In *Arabidopsis,* mutational screens independently identified orthologs of barley RAR1 and SGT1 as components of resistance specified by multiple resistance genes of the CC-NB-LRR type (e.g., *RPM1, RPS2, RPP8*) and TIR-NB-LRR type (e.g., *RPP5, RPS4*), recognizing different bacterial *(Pseudomonas syringae)* and oomycete *(Peronospora parasitica)* pathogens (Austin et al., 2002; Muskett et al., 2002). The *sgt1b* mutation suppressed resistance against *Peronospora parasitica* mediated by the *RPP5* gene, but not as much as the *rpp5* null mutation, and the *rar1* null mutant likewise partially suppressed *RPP5*-mediated resistance (Austin et al., 2002). The *sgt1b/rar1* double mutant exhibited an additive effect of both genes in compromising *RPP5*-mediated resistance with substantially delayed plant HR cell death and whole-cell ROI accumulation (Austin et al., 2002; Muskett et al., 2002). This finding is in agreement with the results obtained by Azevedo and colleagues (2002), who demonstrated a physical interaction between RAR1 and SGT1 and that SGT1 exists in at least two pools (SGT1/RAR1 and SGT1/SCF) with presumably different functions. Collectively, these and additional findings by Tör and colleagues (2002) demonstrated that RAR1 and SGT1 are convergence points of defense signaling conferred by several *R* genes in different plants, and that both have partially combined and distinct roles in resistance, one of which is presumably related to ubiquitination of still-unknown targets.

In addition to ubiquitin, several other ubiquitin-like polypeptide tags such as SUMO (small ubiquitin-like modifiers), RUB (related to ubiquitin), and APG12 (autophagy-defective-12) have been identified in plants. Similar to ubiquitin, these alternative modifiers are attached to -lysyl groups of target proteins, thereby influencing their structure, location, and turnover (Vierstra and Callis, 1999). Evidence for the involvement of the ubiquitin-like modifier SUMO-1 in disease resistance was recently obtained for the interaction between *Nicotiana benthamiana* and the bacterial pathogen *Xanthomonas campestris* (Orth et al., 2000). The bacterial avirulence gene product AvrBsT induces HR cell death and shares significant similarity with the YopJ protein of the human pathogen *Yersinia pestis,* which inhibits the host immune response. These YopJ family members were shown to act as cysteine proteases, specifically removing SUMO-1 residues from its protein conjugates (Orth et al., 2000). Protease-inactive variants of AvrBsT, generated by site-directed mutagenesis, were defective in HR induction. These results indicate the importance of ubiquitin-like protein conjugation and deconjugation in regulation of defense-related signaling pathways leading to HR cell death.

THE ROLE OF CELL ORGANELLES
IN HR INDUCTION

In the animal system, mitochondria represent an important regulatory unit of at least one main pathway leading to PCD. The role of this organelle seems to include sensing and amplification of cellular stress signals arising from other subcellular compartments, such as the nucleus, the endoplasmic reticulum, the cytoskeleton, or the plasma membrane (Ferri and Kroemer, 2001). During the onset of PCD, proteinaceous factors modify the permeability of the outer mitochondrial membrane, which subsequently allows the release of several cell-death activators, including cytochrome c. By interaction with other proteins in the cytosol, cytochrome c can activate the initiator caspase, procaspase-9, thereby finally inducing the ordered disassembly of the cell (Green, 2000). In addition to responding to proteinaceous intracellular death signals, mitochondria may also initiate apoptosis in response to changes in the levels of low molecular weight messengers, such as calcium, to changes in intracellular pH or to changes in the concentration of metabolites reflecting the energy status of the cell, such as ATP, ADP, NADH, etc. (Lam et al., 2001).

Although it is not clear whether cytochrome c leakage occurs during HR cell death in plants, different lines of evidence indicate that plant mitochondria are involved in the regulation of cell death (Lam et al., 2001). For example, the HR-inducing bacterial virulence factor harpin disrupts mitochondrial functions, such as ATP synthesis in tobacco cell cultures or heat dissipation by alternative oxidase (AOX) in *Nicotiana sylvestris* leaves (Xie and Chen, 2000; Boccara et al., 2001). AOX is a protein of the inner mitochondrial membrane, which is exclusively found in plants. It catalyzes an electron flow from the ubiquinol pool to oxygen, thereby creating an electron shunt that bypasses complexes III and IV of the respiratory chain and dissipates energy as heat. Plants deficient in AOX showed increased levels of ROI in the mitochondria and rapid cell death activation concomitant with enhanced expression of defense genes such as PR-1. In contrast, activation or overexpression of AOX suppressed the induction of ROI, the expression of PR-1, and reduced the development of HR symptoms (Lam et al., 2001). These results underscore the importance of the mitochondrial ROI production as one step in intracellular stress signaling.

In addition to mitochondria, several other plant organelles can produce ROI. Photooxidative processes in chloroplasts under high-light conditions and oxidation of the photorespiration product glycolate in peroxisomes are also sources of ROI, indicating that these organelles might also participate in HR-associated cell death signaling. The importance of photooxidative ROI production for HR development is strengthened by the observation that the HR phenotype of several *Arabidopsis* lesion-mimic mutants, including *lsd1* and *lsd3*, is suppressed under short-day conditions (Dietrich et al., 1994). Likewise, catalase-deficient tobacco plants exposed to high light intensities developed necrotic lesions, which may be caused by excess ROI produced by chloroplast metabolism (Chamnongpol et al., 1996). In agreement with the assumption that perturbation of photosynthesis or other chloroplast functions may induce HR-like cell death is the recent characterization of the *Arabidopsis ACD2* gene, which encodes a red chlorophyll catabolite reductase (Mach et al., 2001). It has been suggested that the cell death phenotype of *acd2* mutant plants is caused by the accumulation of chlorophyll breakdown products, which might be specific triggers of cell death or might function by their ability to absorb light and generate free radicals (Mach et al., 2001). Recently, the plastid-localized protein DS9 was shown to modulate the extent and timing of lesion development in tobacco upon infection by tobacco mosaic virus (TMV) (Seo et al., 2000). In plants with reduced levels of the DS9 protein, necrotic spots induced by TMV were smaller, whereas overexpression of DS9 resulted in larger lesions and enhanced systemic spread of the virus, i.e., a reduced level of resistance. DS9 shares significant similarity with the *Escherichia coli* protein FtsH, which can act either as an ATP-dependent metalloprotease or as a chaperone. Because both functions serve as quality-control mechanisms for cellular proteins or protein complexes, these observations indicate that in the absence of DS9, accumulation of misfolded proteins in the chloroplasts may directly or indirectly act as a signal for the induction of HR cell death.

IS HR CELL DEATH REQUIRED FOR RESISTANCE?

Although the HR is a common feature of plant-pathogen interactions and cell death is apparently tightly associated with disease resistance, it is still an open question whether the occurrence of HR is a prerequisite for resistance. Indeed, several examples demonstrate that resistance of plants can be uncoupled from HR development for different types of plant-pathogen interactions, which casts doubt on its direct involvement in disease resistance (Richael and Gilchrist, 1999).

An example demonstrating that *R*-gene-dependent resistance against a virus is not associated with HR cell death is the interaction of potato with potato virus X (PVX). The potato *Rx* gene, a resistance gene of the CC-NB-LRR type, is thought to participate in perception of the PVX coat protein, thereby initiating efficient defense responses. The most striking feature of *Rx*-mediated resistance is the rapid arrest of PVX accumulation in the initially infected cell. However, this inhibition of pathogen spread is not associated with HR cell death (Köhm et al., 1993). The phenomenon was designated extreme resistance, and to elucidate whether this is a particular feature of potato, a transgenic approach using modified plants and viruses was applied. First, transgenic *Nicotiana tabacum* and *N. benthamiana* containing the *Rx* gene exhibited the same rapid arrest of PVX accumulation without HR, demonstrating that HR-independent resistance is not a particular characteristic of potato and can be transferred to heterologous plant species (Bendahmane et al., 1999). Second, HR cell death was triggered by constitutive overexpression of PVX coat protein in tobacco containing the *Rx* gene, indicating that sustained expression of the coat protein elicitor could force extreme resistance to HR resistance (Bendahmane et al., 1999). Finally, to test the relationship between extreme resistance and HR cell death formation as determined by the tobacco *N* gene (TIR-NB-LRR type, conferring resistance to TMV), transgenic TMV expressing both the TMV coat protein and PVX coat proteins of a virulent or an avirulent strain were generated. Tobacco plants containing only the *N* gene showed resistance and typical HR symptoms after inoculation with the recombinant TMV expressing both types of PVX coat protein. HR formation was also observed when *Nicotiana tabacum* lines containing both the *N* gene and the transgenic *Rx* gene were inoculated with the recombinant TMV version carrying the PVX coat protein of the virulent strain. In contrast, inoculation of tobacco lines of the *N* and *Rx* genotype with TMV carrying the PVX coat protein of the avirulent strain resulted in resistance without triggering HR cell death (Bendahmane et al., 1999). Taken together, these results demonstrate that *Rx*-mediated extreme resistance was activated before *N*-mediated onset of HR and that extreme resistance is epistatic to induced HR.

Likewise, *Arabidopsis thaliana* mutants have been derived that display effective gene-for-gene resistance against bacteria without

HR cell death (Bowling et al., 1994; Yu et al., 1998). Screening of a mutagenized *Arabidopsis* line containing the *RPS2* gene by inoculation with a *Pseudomonas syringae* strain expressing *avrRpt2* identified plants mutated at the *DND1* locus. These mutant *dnd1* plants were defective in mounting HR cell death, which is specified by the *avrRpt2/RPS2* gene pair in the wild-type interaction, but they retained characteristic responses to avirulent *Pseudomonas syringae* strains, such as induction of PR proteins and strong restriction of pathogen growth (Yu et al., 1998). They also exhibited enhanced resistance against a broad spectrum of virulent fungal, bacterial, and viral pathogens. The *DND1* gene was recently identified to encode a cyclic nucleotide-gated ion channel, which allows the passage of Ca^{2+}, K^+, and other cations (Clough et al., 2000). This result emphasizes the importance of ion fluxes in plant defense signaling due to physiological changes in membrane permeability that have frequently been observed in plant-pathogen interactions. Since levels of salicylic acid and mRNAs encoding PR proteins are elevated in *dnd1* plants, it has been suggested that this constitutive induction of systemic acquired resistance may substitute for HR cell death in potentiating the stronger defense responses (Yu et al., 1998). Recently, it has been observed that *dnd1* mutant plants also exhibit conditional lesion mimicry (Clough et al., 2000). Therefore, deregulated defense reactions may on the one hand substitute for HR in mounting resistance, but on the other hand may lead to high constitutive levels of protective proteins such as antioxidant enzymes that suppress HR formation by scavenging ROI.

HR-independent resistance of plants to fungal pathogens has also been described. The *Cf-9* gene of tomato mediates resistance to *Cladosporium fulvum,* provided the fungus carries the *avr9* gene (Dangl and Jones, 2001). Resistance in *Cf-9* plants is manifested as gradual cessation of hyphal growth, which occurs exclusively extracellular, without concomitant HR (Hammond-Kosack and Jones, 1994). Interestingly, constitutive expression of recombinant Avr9 protein in transgenic plants containing the *Cf-9* gene was associated with HR cell death, indicating that sustained production or increased amounts of this Avr protein can modify HR-independent resistance such that it becomes associated with HR cell death (Hammond-Kosack et al., 1995).

In barley, *mlo*-mediated resistance to the obligate biotrophic fungus *Blumeria graminis* f. sp. *hordei* is independent of HR cell death. Recessive *mlo* alleles confer broad-spectrum resistance to nearly all *Blumeria* races. At the cytological level, resistance almost invariably leads to the abortion of infection attempts during the penetration process through the epidermal cell wall. Most important, attacked cells do not activate a suicide response and therefore stay alive (Schulze-Lefert and Vogel, 2000). It is presently not known how fungal growth is stopped in *mlo* genotypes, although some evidence suggests that oxidative cross-linking of cell-wall appositions may be involved (von Röpenack et al., 1998). Despite the absence of HR cell death during *mlo*-mediated resistance, there may be a link between resistance and deregulated cell death in *mlo* plants, since one pleiotropic effect of *mlo* alleles is to trigger spontaneous lesions in the absence of the pathogen.

In conclusion, the previous observations indicate that HR is not obligatory for efficient plant defense against pathogens under all circumstances. Two interpretations are possible to explain the lack of such a correlation. First, HR cell death itself might not be important for pathogen restriction but instead be a side effect of other mechanisms or cellular changes which, although limiting pathogen growth, are not compatible with survival of the plant cell. Thus cell death could be the consequence of activated responses such as a strong oxidative burst, ion leakage, or massive synthesis of toxic antimicrobial compounds. Second, HR cell death might be the ultimate plant defense reaction that is mounted only after failure of other defense mechanisms or if the first *R*-gene-mediated signal reaches a certain threshold level. The latter conclusion is supported by the observation that barley plants homozygous for the *Mlg* gene exhibit a characteristic single-cell HR in response to infection by *Blumeria graminis*, whereas plants heterozygous for the *Mlg* gene can still restrict fungal growth but have lost the capacity to trigger HR (Görg et al., 1993).

DO PLANT PATHOGENS MODULATE THE HR IN AN ACTIVE MANNER?

Although HR cell death is not observed in all incompatible plant-pathogen interactions, it seems to contribute to restriction of the invading organism in many cases. As outlined previously by others

(Morel and Dangl, 1997), the impact of the HR may depend on the lifestyle of the pathogen, and therefore it has to be considered whether the parasite grows intra- or extracellularly and whether it is a biotrophic, hemibiotrophic, or necrotrophic pathogen.

Viruses are obligate intracellular parasites that need the host cell machinery for replication. Thus HR cell death of invaded cells appears to be a good measure to block their multiplication; yet not all *R* genes mediating virus resistance are associated with HR development (Morel and Dangl, 1997; Richael and Gilchrist, 1999). Likewise, biotrophic and hemibiotrophic fungal pathogens also depend on nutrient uptake from living host cells, and in all these cases, HR cell death could obviously cause pathogen death. Biotrophic pathogens have therefore evolved mechanisms to actively suppress HR cell death, which is drastically illustrated by the so-called green-island effect caused by virulent pathogens in otherwise senescing leaves (Schulze-Lefert and Vogel, 2000). Indeed, a large number of fungus-derived molecules suppressing plant defense responses have been identified, but their mechanisms of action remain largely unknown (Morel and Dangl, 1997; Heath, 2000).

A sophisticated mechanism of HR modulation by a pathogen has recently been discovered for the interaction between *Arabidopsis thaliana* and *Pseudomonas syringae,* in which the *RPM1/avrRpm1* gene pair specifies resistance. A direct interaction of RPM1, AvrRpm1, and an additional *Arabidopsis* protein (designated RIN4) was demonstrated by co-immunoprecipitation (Mackey et al., 2002). Reduction of RIN4 protein levels by antisense suppression caused diminished RPM1 levels and inhibited both the RPM1-mediated HR and the growth restriction of virulent bacteria carrying *avrRpm1.* Surprisingly, RIN4 reduction resulted in heightened resistance to virulent *Pseudomonas syringae* (lacking *avrRpm1*) and *Peronospora parasitica* pathovars, as well as constitutive expression of typical defense genes. Thus RIN4 functions both as a negative regulator of basal plant defense responses and as a positive regulator of RPM1-mediated resistance. Since *avrRpm1* induces RIN4 phosphorylation, it was suggested that this modification might enhance RIN4 activity as a negative regulator of plant defense, thereby facilitating pathogen growth. In this model, RPM1 would guard against pathogens that use avrRpm1 (functioning as a virulence factor) to manipulate RIN4 activity and thereby suppress the HR and other defense responses.

In contrast to the requirements of biotrophic pathogens which try to prevent HR, necrotrophic pathogens cannot survive only in dead cells, instead they may rather actively induce cell death to promote host invasion and ultimately benefit from the release of nutrients. Indeed, there is ample evidence of pathogens producing phytotoxins that induce HR-like symptoms (Hohn, 1997; Morel and Dangl, 1997). Such toxins may serve as attractive experimental tools to study the role of HR cell death in plant pathogenesis, as was recently demonstrated for fumonisin B1, which is produced by the necrotrophic fungal plant pathogen *Fusarium moniliforme* (Stone et al., 2000). Fumonisin B1 is one of several related mycotoxins produced by some *Fusarium* spp. that elicits an apoptotic form of HR cell death in both plants and animal cell cultures, most probably by inhibition of ceramide synthase, a key enzyme of sphingolipid biosynthesis (Stone et al., 2000). *Arabidopsis* plants infiltrated with fumonisin B1 developed HR-like lesions, concomitant with the generation of ROI and activation of numerous defense responses. These results indicate that *Fusarium moniliforme* and related species may intentionally trigger HR cell death by secretion of fumonisin to obtain nutrients from host tissues.

Some necrotrophic pathogens, such as the fungi *Botrytis cinerea* and *Sclerotinia sclerotiorum,* seem to not only induce but even depend on the plant HR as a prerequisite for rapid host colonization, since their pathogenicity was directly dependent on the level of ROI accumulation in infected plants (Govrin and Levine, 2000). Growth of *B. cinerea* was suppressed in the HR-deficient *Arabidopsis dnd1* mutant (described previously), whereas it was strongly promoted by manipulations that enhanced the concentration of ROI in the plant, such as preinoculation with avirulent bacteria or infiltration of glucose in the presence of glucose oxidase (Govrin and Levine, 2000). In this context, it is interesting to note that necrotrophic fungi were found to contain sets of detoxifying enzymes which allowed them to survive in ROI-rich host environments (Mayer et al., 2001). Likewise, plant pathogenic bacteria also require protective systems against oxidative damage and therefore express ROI scavenging enzymes, such as catalase and superoxide dismutase, which are essential for bacterial viability on some hosts and function as virulence factors (Xu and Pan, 2000; Santos et al., 2001).

DEVELOPMENT OF TRANSGENIC PLANTS: EXPLOITATION OF HR FOR DISEASE CONTROL

With the discovery and isolation of genes that are involved in disease resistance mechanisms in plants, attempts have been made to engineer durable resistance in economically important crop plants (Stuiver and Custers, 2001). Constitutive overexpression of single proteins that are toxic or otherwise interfere with pathogen proliferation, such as viral coat proteins, toxins, enzymes of phytoalexin biosynthesis, chitinases, glucanases, and many other PR proteins, has already been proven successful for enhancing plant resistance (Kombrink and Somssich, 1995, 1997; Stuiver and Custers, 2001) (see Chapter 7). However, a major drawback of modulating resistance by transfer of a single gene is that in many cases broad-spectrum disease control is not provided.

Alternatively, targeted activation of endogenous mechanisms that lead to enhanced resistance might be a useful approach. As outlined earlier, HR cell death is a common and mostly efficient defense system utilized by plants which comprises a complex array of defense responses and signaling mechanisms. Although the HR may stop different kinds of pathogens, including viruses, bacteria, and fungi, it is usually triggered only after specific recognition of pathogen-derived elicitors or Avr proteins. In addition, deregulation of cellular responses located downstream of the initial recognition event, such as formation of ROI, modulation of ion channel activity, or initiation of protein degradation, can also induce HR-like cell death. Based on this knowledge, a number of strategies have been devised to utilize induced HR cell death to engineer resistance.

The "two-component-systems" consist of pathogen-derived *avr* genes or genes encoding elicitors that are introduced and expressed in a plant containing the corresponding recognition system (i.e., R protein or elicitor receptor) to initiate HR cell death. With this approach, successful generation of an HR has been achieved with genes encoding Avr9 from *Cladosporium fulvum*, AvrRpt2 from *Pseudomonas syrin-*

gae, and the elicitor protein cryptogein (elicitin) from *Phytophthora cryptogea* (Hammond-Kosack et al., 1994; McNellis et al., 1998; Keller et al., 1999). An obvious key to the success of this strategy is the selection of an appropriate, tightly regulated promoter, which should be inducible by a variety of pathogens to extend race-specific resistance to broad-spectrum disease resistance. Leakiness of the promoter would result in spontaneous cell death formation with a detrimental influence on plant vigor and yield.

An alternative and more general approach that circumvents the limitations caused by the need for a defined genetic background is to induce cell death by expression of so-called killer genes, encoding products that directly interfere with essential cellular functions. Candidates for such products are RNases, DNases, specific proteases, toxins, etc., several of which have been experimentally evaluated (Mittler and Rizhsky, 2000). For example, expression of the barnase gene, encoding an RNase, under the control of a PR gene promoter resulted in transgenic potato plants with enhanced resistance to *Phytophthora infestans,* supporting the hypothesis that HR cell death at infection sites plays an important role in preventing pathogen proliferation (Strittmatter et al., 1995). However, growth under greenhouse or field conditions ultimately led to self-destruction of the plants, indicative of an endogenous activation of the transgene in aging plants, which underscores the need for specific, pathogen-responsive promoters. Correspondingly, identification and isolation of such selectively activated and tightly regulated promoters or *cis*-acting promoter elements is presently an intensively studied area (Keller et al., 1999; Pontier et al., 2001; Rushton et al., 2002).

Other transgenes inducing lesion-mimic phenotypes encode components of downstream signaling pathways involved in HR development or compounds that activate or interfere with their function. Thus compounds that mimic ion fluxes across the plasma membrane, such as the bacterio-opsin, a bacterial proton pump, or cholera toxin, an inhibitor of GTPase and G-protein signaling, both induce HR-like cell death, which is correlated with PR gene expression and elevated disease resistance (Mittler and Rizhsky, 2000). Likewise, expression of metabolic enzymes that either generate peroxides (e.g., glucose oxidase) or antisense suppression of those that catalyze their detoxification (e.g., catalase, ascorbate peroxidase) was also found to induce the formation of lesions, expression of defense-related genes and enhanced resistance (Wu et al., 1995; Chamnongpol et al., 1996; Mittler and Rizhsky, 2000). These results clearly impress upon the tight connection between enhanced ROI production and HR cell death. Manipulation of ubiquitin-regulated protein degradation pathways can also result in lesion-mimic phenotypes (Becker et al., 1993). Transformation of tobacco with a modified ubiquitin that is unable to polymerize an essential step in the ubiquitin-degradation pathway led to spontaneous lesion formation; however, challenging these plants with TMV resulted in fewer lesions and reduced virus replication in comparison to control plants.

Although artificial generation of HR seems a promising approach to control biotrophic and hemibiotrophic pathogens, it may be counterproductive with respect to necrotrophic pathogens, which apparently depend on dead host cells for growth. Accordingly, suppression of HR cell death appears to be an appropriate measure to limit these types of pathogens. Experimental evidence for the validity of this concept has recently been obtained. Transgenic tobacco plants expressing negative regulators of apoptosis, such as the human *Bcl-2*, human *Bcl-X$_L$*, or nematode *Ced-9* genes, exhibited resistance to several necrotrophic fungal pathogens, including *Sclerotinia sclerotiorum, Botrytis cinerea,* and *Cercospora nicotianae,* as well as to tomato spotted wilt virus (Dickman et al., 2001). Plants harboring *Bcl-X$_L$* with a loss-of-function point mutation did not protect against pathogen challenge, demonstrating that resistance was dependent on a functional transgene and was not due to unspecific stress caused by the heterologous gene product (Dickman et al., 2001). These results further suggest that cellular pathways and mechanisms for cell death control are conserved between animals and plants.

CONCLUSIONS AND FUTURE DIRECTIONS

The molecular and biochemical events that occur at the sites of infection in plants include a plethora of changes triggered by both host and pathogen genes/factors. Among these, the induction of HR cell death seems to be the most common response to many different kinds of biotic and abiotic stresses. However, the functional role of the HR in limiting growth of various types of pathogens is not fully under-stood. Recent work on HR cell death in plants has uncovered an increasing number of morphological and biochemical features that are identical or similar to cellular events occurring during apoptosis in animal systems. This may indicate that the ordered dismantling of the cell is an ancient mechanism that has been conserved during evolution and that is utilized during developmental processes as well as during the cellular immune response across kingdoms. However, despite obvious similarities between plant HR cell death and animal apoptosis, differences are also apparent. In particular, it is at present unresolved to what extent signaling pathways overlap and control elements (regulators) of HR and PCD are structurally and functionally conserved. Thus one of the future challenges will be to unravel the complete signaling network that is engaged in cell death control. From the current information, it is clear that R-protein-mediated pathogen recognition and execution of HR cell death are connected via a complex array of signaling components, involving parallel pathways, branching and convergence points, positive and negative regulatory components, small molecules, large protein complexes, and even whole cell organelles. Integrated research efforts that combine genetics, biochemistry, and cell biology, complemented by bioinformatics, will be required to overcome the enduring challenges and mysteries related to HR cell death and the mechanisms of its control and execution.

REFERENCES

Aravind, L., Dixit, V.M., and Koonin, E.V. (1999). The domains of death: Evolution of the apoptosis machinery. *Trends in Biochemical Sciences* 24: 47-53.

Austin, M.J., Muskett, P., Kahn, K., Feys, B.J., Jones, J.D.G., and Parker, J.E. (2002). Regulatory role of *SGT1* in early *R* gene-mediated plant defenses. *Science* 295: 2077-2080.

Azevedo, C., Sadanandom, A., Kitagawa, K., Freialdenhoven, A., Shirasu, K., and Schulze-Lefert, P. (2002). The RAR1 interactor SGT1, an essential component of *R* gene-triggered disease resistance. *Science* 295: 2073-2076.

Becker, F., Buschfeld, E., Schell, J., and Bachmaier, A. (1993). Altered response to viral infection by tobacco plants perturbed in ubiquitin system. *The Plant Journal* 3: 875-881.

Bendahmane, A., Kanyuka, K., and Baulcombe, D.C. (1999). The Rx gene from potato controls separate virus resistance and cell death responses. *The Plant Cell* 11: 781-791.

Boccara, M., Boué, C., Garmier, M., De Paepe, R., and Boccara, A.-C. (2001). Infra-red thermography revealed a role for mitochondria in pre-symptomatic cooling during harpin-induced hypersensitive response. *The Plant Journal* 28: 663-670.

Bowling, S.A., Guo, A., Cao, H., Gordon, A.S., Klessig, D.F., and Dong, X. (1994). A mutation in *Arabidopsis* that leads to constitutive expression of systemic acquired resistance. *The Plant Cell* 6: 1845-1857.

Boyes, D.C., Nam, J., and Dangl, J.L. (1998). The *Arabidopsis thaliana* RPM1 disease resistance gene product is a peripheral plasma membrane protein that is degraded coincident with the hypersensitive response. *Proceedings of the National Academy of Sciences, USA* 95: 15849-15854.

Büschges, R., Hollricher, K., Panstruga, R., Simons, G., Wolter, M., Frijters, A., van Daelen, R., van der Lee, T., Diergaarde, P., Groenendijk, J., et al. (1997). The barley *Mlo* gene: A novel control element of plant pathogen resistance. *Cell* 88: 695-705.

Callis, J. and Vierstra, R.D. (2000). Protein degradation in signaling. *Current Opinion in Plant Biology* 3: 381-386.

Century, K.S., Holub, E.B., and Staskawicz, B.J. (1995). *NDR1*, a locus of *Arabidopsis thaliana* that is required for disease resistance to both a bacterial and fungal pathogen. *Proceedings of the National Academy of Sciences, USA* 92: 6597-6601.

Century, K.S., Shapiro, A.D., Repetti, P.P., Dahlbeck, D., Holub, E., and Staskawicz, B.J. (1997). *NDR1*, a pathogen-induced component required for *Arabidopsis* disease resistance. *Science* 278: 1963-1965.

Chamnongpol, S., Willekens, H., Langebartels, C., van Montagu, M., Inzé, D., and Van Camp, W. (1996). Transgenic tobacco with a reduced catalase activity develop necrotic lesions and induces pathogenesis-related expression under high light. *The Plant Journal* 10: 491-503.

Clough, S.J., Fengler, K.A., Yu, I.-C., Lippok, B., Smith, R.K. Jr., and Bent, A.F. (2000). The *Arabidopsis dnd1* "defense, no death" gene encodes a mutated cyclic nucleotide-gated ion channel. *Proceedings of the National Academy of Sciences, USA* 97: 9323-9328.

Dangl, J.L. and Jones, J.D.G. (2001). Plant pathogens and integrated defence responses to infection. *Nature* 411: 826-833.

Danon, A., Delorme, V., Mailhac, N., and Gallois, P. (2000). Plant programmed cell death: A common way to die. *Plant Physiology and Biochemistry* 38: 647-655.

Delledonne, M., Zeier, J., Marocco, A., and Lamb, C. (2001). Signal interactions between nitric oxide and reactive oxygen intermediates in the plant hypersensitive disease resistance response. *Proceedings of the National Academy of Sciences, USA* 98: 13454-13459.

Devoto, A., Piffanelli, P., Nilsson, I.M., Wallin, E., Panstruga, R., von Heijne, G., and Schulze-Lefert, P. (1999). Topology, subcellular localization, and sequence diversity of the Mlo family in plants. *The Journal of Biological Chemistry* 274: 34993-35005.

Dickman, M.B., Park, Y.K., Oltersdorf, T., Li, W., Clemente, T., and French, R. (2001). Abrogation of disease development in plants expressing animal anapoptotic genes. *Proceedings of the National Academy of Sciences, USA* 98: 6957-6962.

Dietrich, R.A., Delaney, T.P., Uknes, S.J., Ward, E.R., Ryals, J.A., and Dangl, J.L. (1994). *Arabidopsis* mutants simulating disease resistance response. *Cell* 77: 565-577.

Dietrich, R.A., Richberg, M.H., Schmidt, R., Dean, C., and Dangl, J.L. (1997). A novel zinc finger protein is encoded by the *Arabidopsis LSD1* gene and functions as a negative regulator of plant cell death. *Cell* 88: 685-694.

Doke, N. (1983). Involvement of superoxide anion generation in the hypersensitive response of potato-tuber tissues to infection with an incompatible race of *Phytophthora infestans* and to the hyphal wall components. *Physiological Plant Pathology* 23: 345-357.

Draper, J. (1997). Salicylate, superoxide synthesis and cell suicide in plant defense. *Trends in Plant Science* 2: 162-165.

Durner, J., Wendehenne, D., and Klessig, D.F. (1998). Defense gene induction in tobacco by nitric oxide, cyclic GMP, and cyclic ADP-ribose. *Proceedings of the National Academy of Sciences, USA* 95: 10328-10333.

Estelle, M. (2001). Proteases and cellular regulation in plants. *Current Opinion in Plant Biology* 4: 254-260.

Falk, A., Feys, B.J., Frost, L.N., Jones, J.D.G., Daniels, M.J., and Parker, J.E. (1999). *EDS1*, an essential component of *R* gene-mediated disease resistance in *Arabidopsis* has homology to eukaryotic lipases. *Proceedings of the National Academy of Sciences, USA* 96: 3292-3297.

Ferri, K.F. and Kroemer, G. (2001). Mitochondria—The suicide organelles. *BioEssays* 23: 111-115.

Feys, B.J., Moisan, L.J., Newman, M.-A., and Parker, J.E. (2001). Direct interaction between the *Arabidopsis* disease resistance signaling proteins, EDS1 and PAD4. *The EMBO Journal* 20: 5400-5411.

Feys, B.J. and Parker, J.E. (2000). Interplay of signaling pathways in plant disease resistance. *Trends in Genetics* 16: 449-455.

Frye, C.A., Tang, D., and Innes, R.W. (2001). Negative regulation of defense responses in plants by a conserved MAPKK kinase. *Proceedings of the National Academy of Sciences, USA* 98: 373-378.

Glazebrook, J. (2001). Genes controlling expression of defense responses in *Arabidopsis*—2001 status. *Current Opinion in Plant Biology* 4: 301-308.

Görg, R., Hollricher, K., and Schulze-Lefert, P. (1993). Functional analysis and RFLP-mediated mapping of the *Mlg* resistance locus in barley. *The Plant Journal* 3: 857-866.

Govrin, E.M. and Levine, A. (2000). The hypersensitive response facilitates plant infection by the necrotrophic pathogen *Botrytis cinerea*. *Current Biology* 10: 751-757.

Grant, J.J. and Loake, G.J. (2000). Role of reactive oxygen intermediates and cognate redox signaling in disease resistance. *Plant Physiology* 124: 21-29.

Green, D.R. (2000). Apoptotic pathways: Paper wraps stone blunts scissors. *Cell* 102: 1-4.

Grütter, M.G. (2000). Caspases: Key players in programmed cell death. *Current Opinion in Structural Biology* 10: 649-655.

Halterman, D., Zhou, F., Wei, F., Wise, R.P., and Schulze-Lefert, P. (2001). The MLA6 coiled-coil, NBS-RLL protein confers *AvrMla6*-dependent resistance specificity to *Blumeria graminis* f. sp. *hordei* in barley and wheat. *The Plant Journal* 25: 335-348.

Hammond-Kosack, K.E., Harrison, K., and Jones, J.D.G. (1994). Developmentally regulated cell death on expression of the fungal avirulence gene *Avr9* in tomato seedlings carrying the disease-resistance gene *Cf-9*. *Proceedings of the National Academy of Sciences, USA* 91: 10445-10449.

Hammond-Kosack, K.E. and Jones, J.D.G. (1994). Incomplete dominance of tomato *Cf* genes for resistance to *Cladosporium fulvum*. *Molecular Plant-Microbe Interactions* 7: 58-70.

Hammond-Kosack, K.E. and Jones, J.D.G. (1996). Resistance gene-dependent plant defense responses. *The Plant Cell* 8: 1773-1791.

Hammond-Kosack, K.E., Staskawicz, B.J., Jones, J.D.G., and Baulcombe, D.C. (1995). Functional expression of a fungal avirulence gene from a modified potato virus X genome. *Molecular Plant-Microbe Interactions* 8: 181-185.

He, Z., Wang, Z.Y., Li, J., Zhu, Q., Lamb, C., Ronald, P., and Chory, J. (2000). Perception of brassinosteroids by the extracellular domain of the receptor kinase BRI1. *Science* 288: 2360-2363.

Heath, M.C. (1998). Apoptosis, programmed cell death and the hypersensitive response. *European Journal of Plant Pathology* 104: 117-124.

Heath, M.C. (2000). Hypersensitive response-related death. *Plant Molecular Biology* 44: 321-334.

Heo, W.D., Lee, S.H., Kim, M.C., Kim, J.C., Chung, W.S., Chun, H.J., Lee, K.J., Park, C.Y., Park, H.C., Choi, J.Y., and Cho, M.J. (1999). Involvement of specific calmodulin isoforms in salicylic acid-independent activation of plant disease resistance responses. *Proceedings of the National Academy of Sciences, USA* 96: 766-771.

Hohn, T.M. (1997). Fungal phytotoxins: Biosynthesis and activity. In Carroll, G. and Tudzynski, P. (Eds.), *The Mycota*, Volume 5, Part A, *Plant Relationships* (pp. 127-144). Berlin, Heidelberg: Springer-Verlag.

Hückelhoven, R., Fodor, J., Preis, C., and Kogel, K.-H. (1999). Hypersensitive cell death and papilla formation in barley attacked by the powdery mildew fungus are associated with hydrogen peroxide but not with salicylic acid accumulation. *Plant Physiology* 119: 1251-1260.

Jabs, T., Tschöpe, M., Colling, C., Hahlbrock, K., and Scheel, D. (1997). Elicitor-stimulated ion fluxes and O_2^- from the oxidative burst are essential components in triggering defense gene activation and phytoalexin synthesis in parsley. *Proceedings of the National Academy of Sciences, USA* 94: 4800-4805.

Jones, J.D.G. (2001). Putting knowledge of plant disease resistance genes to work. *Current Opinion in Plant Biology* 4: 281-287.

Jørgensen, J.H. (1996). Effect of three suppressors on the expression of powdery mildew resistance genes in barley. *Genome* 39: 492-498.

Karniol, B. and Chamovitz, D.A. (2000). The COP9 signalosome: From light signaling to general developmental regulation and back. *Current Opinion in Plant Biology* 3: 387-393.

Keller, H., Pamboukdjian, N., Ponchet, M., Poupet, A., Delong, R., Verrier, J.-L., Roby, D., and Ricci, P. (1999). Pathogen-induced elicitin production in transgenic tobacco generates a hypersensitive response and nonspecific disease resistance. *The Plant Cell* 11: 223-235.

Kim, M.C., Panstruga, R., Elliott, C., Müller, J., Devoto, A., Yoon, H.W., Park, H.C., Cho, M.J., and Schulze-Lefert, P. (2002). Calmodulin interacts with MLO protein to regulate defence against mildew in barley. *Nature* 416: 447-450.

Kitagawa, K., Skowyra, D., Elledge, S.J., Harper, J.W., and Hieter, P. (1999). SGT1 encodes an essential component of the yeast kinetochore assembly pathway and a novel subunit of the SCF ubiquitin complex. *Molecular Cell* 4: 21-33.

Kjemtrup, S., Nimchuk, Z., and Dangl, J.L. (2000). Effector proteins of phytopathogenic bacteria: Bifunctional signals in virulence and host recognition. *Current Opinion in Microbiology* 3: 73-78.

Klessig, D.F., Durner, J., Noad, R., Navarre, D.A., Wendehenne, D., Kumar, D., Zhou, J.M., Shah, J., Zhang, S., Kachroo, P., et al. (2000). Nitric oxide and salicylic acid signalling in plant defense. *Proceedings of the National Academy of Sciences, USA* 97: 8849-8855.

Kliebenstein, D.J., Dietrich, R.A., Martin, A.C., Last, R.L., and Dangl, J.L. (1999). LSD1 regulates salicylic acid induction of copper zinc superoxide dismutase in *Arabidopsis thaliana. Molecular Plant-Microbe Interactions* 12: 1022-1026.

Köhm, B.A., Goulden, M.G., Gilbert, J.E., Kavanagh, T.A., and Baulcombe, D.C. (1993). A potato virus X resistance gene mediates an induced, nonspecific resistance in protoplasts. *The Plant Cell* 5: 913-920.

Kombrink, E. and Somssich, I. E. (1995). Defense responses of plants to pathogens. *Advances in Botanical Research* (formerly *Advances in Plant Pathology*) 21: 1-34.

Kombrink, E. and Somssich, I.E. (1997). Pathogenesis-related proteins and plant defense. In Carroll, G. and Tudzynski, P. (Eds.), *The Mycota*, Volume 5, Part A, *Plant Relationships* (pp. 107-128). Berlin, Heidelberg: Springer-Verlag.

Korthout, H.A.A.J., Berecki, G., Bruin, W., van Duijn, B., and Wang, M. (2000). The presence and subcellular localization of caspase 3-like proteinases in plant cells. *FEBS Letters* 16: 139-144.

Lam, E. and del Pozo, O. (2000). Caspase-like protease involvement in the control of plant cell death. *Plant Molecular Biology* 44: 417-428.

Lam, E., Kato, N., and Lawton, M.A. (2001). Programmed cell death, mitochondria and the plant hypersensitive response. *Nature* 411: 848-853.

Lamb, C. and Dixon, R.A. (1997). The oxidative burst in plant disease resistance. *Annual Review of Plant Physiology and Plant Molecular Biology* 48: 251-275.

Levine, A., Pennell, R.I., Alvarez, M.E., Palmer, R., and Lamb, C. (1996). Calcium-mediated apoptosis in plant hypersensitive disease resistance response. *Current Biology* 6: 427-437.

Levine, A., Tenhaken, R., Dixon, R., and Lamb, C. (1994). H_2O_2 from the oxidative burst orchestrates the plant hypersensitive disease resistance response. *Cell* 79: 583-593.

Mach, J.M., Castillo, A.R., Hoogstraten, R., and Greenberg, J.T. (2001). The *Arabidopsis*-accelerated cell death gene *ACD2* encodes red chlorophyll catabolite reductase and suppresses the spread of disease symptoms. *Proceedings of the National Academy of Sciences, USA* 98: 771-776.

Mackey, D., Holt B.F., III, Wiig, A., and Dangl, J.L. (2002). RIN4 interacts with *Pseudomonas syringae* type III effector molecules and is required for RPM1-mediated resistance in *Arabidopsis*. *Cell* 108: 743-754.

Mayer, A.M., Staples, R.C., and Gil-ad, N.L. (2001). Mechanisms of survival of necrotrophic fungal pathogens in hosts expressing the hypersensitive response. *Phytochemistry* 58: 33-41.

McNellis, T.W., Mudgett, M.B., Li, K., Aoyama, T., Horvath, D., Chua, N.-H., and Staskawicz, B.J. (1998). Glucocorticoid-inducible expression of a bacterial avirulence gene in transgenic *Arabidopsis* induces hypersensitive cell death. *The Plant Journal* 14: 247-257.

Mittler, R. and Rizhsky, L. (2000). Transgene-induced lesion mimic. *Plant Molecular Biology* 44: 335-344.

Morel, J.-B. and Dangl, J. (1997). The hypersensitive response and the induction of cell death in plants. *Cell Death and Differentiation* 4: 671-683.

Muskett, P.R., Kahn, K., Austin, M.J., Moisan, L.J., Sadanandom, A., Shirasu, K., Jones, J. D.G., and Parker, J.E. (2002). *Arabidopsis RAR1* exerts rate-limiting control of *R* gene-mediated defenses against multiple pathogens. *The Plant Cell* 14: 979-992.

Nürnberger, T. and Scheel, D. (2001). Signal transduction in the plant immune response. *Trends in Plant Science* 6: 372-379.

Orth, K., Xu, Z., Mudgett, M.B., Bao, Z.Q., Palmer, L.E., Bliska, J.B., Mangel, W.F., Staskawicz, B., and Dixon, J.E. (2000). Disruption of signaling by *Yersinia* effector YopJ, a ubiquitin-like protein protease. *Science* 290: 1594-1597.

Pennell, R.I. and Lamb, C. (1997). Programmed cell death in plants. *The Plant Cell* 9: 1157-1168.

Petersen, M., Brodersen, P., Naested, H., Andreasson, E., Linhart, U., Johansen, B., Nielsen, H.B., Lacy, M., Austin, M.J., Parker, J.E., et al. (2000). *Arabidopsis* MAP kinase 4 negatively regulates systemic acquired resistance. *Cell* 103: 1111-1120.

Pontier, D., Balagué, C., Bezombes-Marion, I., Tronchet, M., Deslandes, L., and Roby, D. (2001). Identification of a novel pathogen-responsive element in the promoter of the tobacco gene *HSR203J*, a molecular marker of the hypersensitive response. *The Plant Journal* 26: 495-507.

Rate, D.N., Cuenca, J.V., Bowman, G.R., Guttman, D.S., and Greenberg, J.T. (1999). The gain-of-function *Arabidopsis acd6* mutant reveals novel regulation and function of the salicylic acid signaling pathway in controlling cell death, defenses, and cell growth. *The Plant Cell* 11: 1695-1708.

Richael, C. and Gilchrist, D. (1999). The hypersensitive response: A case of hold or fold? *Physiological and Molecular Plant Pathology* 55: 5-12.

Romeis, T. (2001). Protein kinases in the plant defence response. *Current Opinion in Plant Biology* 4: 407-414.

Rushton, P.J., Reinstädler, A., Lipka, V., Lippok, B., and Somssich, I.E. (2002). Synthetic plant promoters containing defined regulatory elements provide novel insights into pathogen- and wound-induced signaling. *The Plant Cell* 14: 749-762.

Rustérucci, C., Aviv, D.H., Hold B.F. III, Dangl, J.L., and Parker, J.E. (2001). The disease resistance signaling components *EDS1* and *PAD4* are essential regulators of the cell death pathway controlled by *LSD1* in *Arabidopsis*. *The Plant Cell* 13: 2211-2224.

Santos, R., Franza, T., Laporte, M.L., Sauvage, C., Touati, D., and Expert, D. (2001). Essential role of superoxide dismutase on the pathogenicity of *Erwinia chrysanthemi* strain 3937. *Molecular Plant-Microbe Interactions* 14: 758-767.

Schulze-Lefert, P. and Vogel, J. (2000). Closing the ranks to attack by powdery mildew. *Trends in Plant Science* 5: 343-348.

Seo, S., Okamoto, M., Iwai, T., Iwano, M., Fukui, K., Isogai, A., Nakajima, N., and Ohashi, Y. (2000). Reduced levels of chloroplast FtsH protein in tobacco mosaic virus-infected tobacco leaves accelerate the hypersensitive reaction. *The Plant Cell* 12: 917-932.

Shah, J., Kachroo, P., and Klessig, D.F. (1999). The *Arabidopsis ssi* mutation restores pathogenesis-related gene expression in *npr1* plants and renders defensin gene expression salicylic acid dependent. *The Plant Cell* 11: 191-206.

Shirasu, K., Lahaye, T., Tan, M.-W., Zhou, F., Azevedo, C., and Schulze-Lefert, P. (1999). A novel class of eukaryotic zinc-binding proteins is required for disease resistance signaling in barley and development in *C. elegans*. *Cell* 99: 355-366.

Shirasu, K. and Schulze-Lefert, P. (2000). Regulators of cell death in disease resistance. *Plant Molecular Biology* 44: 371-385.

Smirnoff, N. (2000). Ascorbic acid: Metabolism and functions of a multi-facetted molecule. *Current Opinion in Plant Biology* 3: 229-235.

Stakman, E.C. (1915). Relation between *Puccinia graminis* and plants highly resistant to its attack. *Journal of Agricultural Research* 4: 193-199.

Sticher, L., Mauch-Mani, B., and Métraux, J.P. (1997). Systemic acquired resistance. *Annual Review of Phytopathology* 35: 235-270.

Stone, J.M., Heard, J.E., Asai, T., and Ausubel, F.M. (2000). Simulation of fugal-mediated cell death by fumonisin B1 and selection of fumonisin B1-resistant *(fbr) Arabidopsis* mutants. *The Plant Cell* 12: 1811-1822.

Strittmatter, G., Janssnes, J., Opsomer, C., and Botterman, J. (1995). Inhibition of fungal disease development in plants by engineering controlled cell death. *Bio/Technology* 13: 1085-1089.

Stuiver, M.H. and Custers, J.H.H.V. (2001). Engineering disease resistance in plants. *Nature* 411: 865-868.

Suzuki, N., Suzuki, S., Duncan, G.S., Millar, D.G., Wada, T., Mirtsos, C., Takada, H., Wakeham, A., Itie, A., Li, S., et al. (2002). Severe impairment of interleukin-1 and Toll-like signalling in mice lacking IRAK-4. *Nature* 416: 750-754.

Tang, X., Frederick, R.D., Zhou, J., Halterman, D.A., Jia, Y., and Martin, G.B. (1996). Initiation of plant disease resistance by physical interaction of AvrPto and Pto kinase. *Science* 274: 2060-2063.

Tör, M., Gordon, P., Cuzick, A., Eulgem, T., Sinapidou, E., Mert-Türk, F., Can, C., Dangl, J. L., and Holub, E.B. (2002). *Arabidopsis* SGT1b is required for defense signaling conferred by several downy mildew resistance genes. *The Plant Cell* 14: 993-1003.

Vierstra, R.D. and Callis, J. (1999). Polypeptide tags, ubiquitous modifiers for plant protein regulation. *Plant Molecular Biology* 41: 435-442.

von Röpenack, E., Parr, A., and Schulze-Lefert, P. (1998). Structural analyses and dynamics of soluble and cell wall-bound phenolics in a broad spectrum resistance to the powdery mildew fungus in barley. *Journal of Biological Chemistry* 273: 9013-9022.

Weymann, K., Hunt, M., Uknes, S., Neuenschwander, U., Lawton, K., Steiner, H.Y., and Ryals, J. (1995). Suppression and restoration of lesion formation in *Arabidopsis lsd* mutants. *The Plant Cell* 7: 2013-2022.

Wojtaszek, P. (1997). Oxidative burst: An early plant response to pathogen infection. *Biochemical Journal* 322: 681-692.

Wu, G., Shortt, B.J., Lawrence, E.B., Levine, E.B., Fitzsimmons, K.C., and Shah, D.M. (1995). Disease resistance conferred by expression of a gene encoding H_2O_2-generating glucose oxidase in transgenic potato plants. *The Plant Cell* 7: 1357-1368.

Xie, Z. and Chen, Z. (2000). Harpin-induced hypersensitive cell death is associated with altered mitochondrial functions in tobacco cells. *Molecular Plant-Microbe Interactions* 13: 183-190.

Xu, X.Q. and Pan, S.Q. (2000). An *Agrobacterium* catalase is a virulence factor involved in tumorigenesis. *Molecular Microbiology* 35: 407-414.

Yu, I.-C., Parker, J., and Bent, A.F. (1998). Gene-for-gene disease resistance without the hypersensitive response in *Arabidopsis dnd1* mutant. *Proceedings of the National Academy of Sciences, USA* 95: 7819-7824.

Zhou, F., Kurth, J., Wei, F., Elliott, C., Valè, G., Yahiaoui, N., Keller, B., Somerville, S., Wise, R., and Schulze-Lefert, P. (2001). Cell-autonomous expression of barley *Mla1* confers race-specific resistance to the powdery mildew fungus via a *Rar1*-independent signaling pathway. *The Plant Cell* 13: 337-350.

Chapter 4

Fungal (A)Virulence Factors at the Crossroads of Disease Susceptibility and Resistance

N. Westerink
M. H. A. J. Joosten
P. J. G. M. de Wit

INTRODUCTION

Pathogenic fungi use diverse strategies to ingress their host plants. Some pathogens enter plants through wounds or natural openings, whereas others use specialized structures, such as appressoria, to penetrate intact plant surfaces or enter the host using cuticle- and cell wall–degrading enzymes. Most fungal pathogens colonize all plant organs, such as leaves, stems, and roots, either by growing between the cells as intercellular mycelium or by penetrating the cells and subsequently growing as intracellular mycelium. Some fungi kill their host and feed on dead tissue (necrotrophs), while others colonize the living host (biotrophs) or even require living tissue to complete their life cycle (obligates). During the biotrophic phase, signal and nutrient exchange between pathogen and host is often mediated by specialized infection structures, such as haustoria.

We thank H. A. Van den Burg, C. F. De Jong, and R. Luderer for making available unpublished work and R. A. L. Van der Hoorn for drawing the figures. N. Westerink is supported by research grants from Wageningen University. The research team of M. H. A. J. Joosten and P. J. G. M. de Wit, working on the interaction between *C. fulvum* and tomato at the Laboratory of Phytopathology at Wageningen University, is supported by NWO-ALW (Netherlands Organisation for Scientific Research, Council Earth and Life Science) and NWO-STW (Netherlands Organisation for Scientific Research, Technology Foundation).

Most plants are resistant toward the majority of pathogenic fungi. A common and effective durable type of resistance is nonhost resistance that prevents plants from becoming infected by potential pathogens. Nonhost resistance often involves a protection provided by physical barriers or by early signaling events and highly localized responses within the cell wall (Heath, 2000). Host resistance, however, is usually restricted to a particular pathogen species and is commonly expressed against specific pathogen genotypes. In this case, the plant specifically recognizes the invading pathogen and active defense responses are induced that lead to resistance. Elicitation of defense responses is mediated by the perception of pathogen signal molecules encoded by avirulence *(Avr)* genes only when the matching plant resistance *(R)* gene is present, which results in an incompatible interaction between host (resistant) and pathogen (avirulent). If the *R* and/or *Avr* gene is absent or nonfunctional, the interaction between host (susceptible) and pathogen (virulent) is compatible. As opposed to the basal defense responses that often partially inhibit pathogens during colonization of the host plant, *R*-gene-mediated resistance involves a rapid and effective defense mechanism that is often associated with a localized death of plant cells, called the hypersensitive response (HR) (see Chapter 3).

As opposed to race-specific elicitors encoded by *Avr* genes, race-nonspecific (or general) elicitors stimulate defense responses in all genotypes of at least one plant species. These general elicitors are often indirect or not direct products of *Avr* genes, but rather structural fungal cell wall components (such as chitin- or glucan-oligosaccharides) released by plant hydrolytic enzymes (Nürnberger, 1999). In this chapter, we will focus on fungal *Avr* gene products (race-specific elicitors) that confer species- or genotype-specific resistance. The function of *Avr* genes as avirulence determinants—how *Avr* gene products induce *R*-gene-mediated resistance, as well as virulence determinants, i.e., how *Avr* gene products contribute to virulence of the pathogen—will be discussed in detail. Four models are presented that illustrate different mechanisms underlying perception of *Avr* gene products by plants, leading either to disease susceptibility or resistance.

BACKGROUND

The first report that described resistance of plants to fungal pathogens goes back to the end of the nineteenth century, where Farrer showed that certain wheat cultivars were resistant to the rust fungus *Puccinia graminis* f. sp. *tritici* (Farrer, 1898). A few years later, in 1905, Biffen reported that wheat varieties and their progeny inherited resistance toward *Puccinia striiformis* in a Mendelian fashion (Biffen, 1905). In subsequent years, studies revealed that the resistance character was often a dominant monogenic trait, which provided the possibility to breed for resistance against pathogens. Soon after the introduction of resistant plants in agriculture, however, varieties that were initially resistant to a given pathogen subsequently became infected. In all cases, the changes were due to the appearance of new physiological races of the pathogen that were able to overcome resistance. The genetic basis of variability within a pathogen species was first described by Johnson, who crossed two races of *P. graminis* f. sp. *tritici* and showed that inheritance of (a)virulence also followed Mendel's law (Johnson et al., 1934). Flor, working on the *Melampsora lini*-flax interaction, and Oort, working on the *Ustilago tritici*-wheat interaction, were the first to present the genetic basis of specific gene-for-gene interactions between a host plant and a pathogen (Flor, 1942; Oort, 1944). These authors demonstrated that (a)virulence of physiologic races of *M. lini* and *U. tritici* was conditioned by a single pair of genes specific for each host-pathogen interaction. This gene-for-gene relationship refers to an interaction, whereby for each dominant resistance gene in the host there is a corresponding avirulence gene in the pathogen. By crossing different races of *M. lini* that were virulent on a particular flax variety with races that were avirulent, Flor showed that avirulence and virulence of pathogens was inherited as a dominant and as a recessive trait, respectively (Flor, 1958). At that time, the nature of the "mutations" leading to virulence in the flax rust fungus was unknown. Day (1957) postulated that "changing the parasite substance taking part in the primary interaction between host and pathogen would abolish defence responses leading to plant disease resistance" (pp. 1141-1142) Indeed, recent genetic and biochemical data obtained from various host-pathogen interactions for which a gene-for-gene relationship has been de-

scribed, and which involve either viruses, bacteria, fungi, or nematodes, reveal that elicitation of defense responses is circumvented by
mutations or deletions in an *Avr* gene (Nürnberger, 1999).

To explain the molecular basis of the gene-for-gene concept, various models have been proposed, which will be discussed in detail
(see Figure 4.1). Consistent with all models is that the product of the
Avr gene is recognized, either directly or indirectly, by the product of

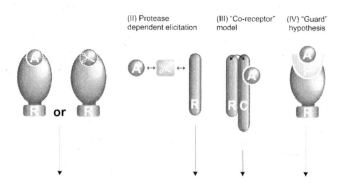

FIGURE 4.1. Schematic representation of gene-for-gene interactions at the protein level. Four models have been proposed that describe either direct (I) or indirect (II-IV) perception of avirulence (AVR) proteins by plant-resistance (R)
proteins. (I) This classical model predicts a direct interaction between an AVR
protein (A) and a matching R protein. Defense responses are induced independently or dependently (as illustrated by the scissors), of the proteolytic activity of
the AVR protein. (II) Binding of AVR protein to host-encoded proteins might
involve the generation of protease-dependent elicitor peptide(s) or complex(es).
The AVR protein, on the other hand, might also trigger *R* gene-mediated resistance by suppressing the generation of proteolytically processed negative regulators of defense responses. (III) The AVR protein binds to, at least, an additional
component (C), which subsequently interacts with the R protein to trigger
defense responses. The interaction between R protein and "coreceptor" might
either be required for receptor activation after AVR binding or for recruitment of a
functional receptor complex that mediates AVR recognition. (IV) According to
the "guard" hypothesis, the R protein safeguards the virulence target (V) of the
AVR protein. For further details, see text. (*Source:* Adapted from Bonas and
Lahaye, 2002).

the corresponding *R* gene present in the resistant plant. This recognition is often associated with a rapid local necrosis of host cells at the site of penetration, the hypersensitive response, which is the hallmark of gene-for-gene-based resistance and resembles programmed cell death in animals (for details, see Chapter 3). The HR is associated with the induction of defense-related responses, including lignification, cell wall enforcement, callose deposition, accumulation of phytoalexins, and transcription of genes encoding pathogenesis-related (PR) proteins that prevent further spread of the invading pathogen (for details, see Chapters 1, 2, and 5). To date, a variety of *Avr* genes have been identified which encode proteins that trigger defense responses in plants carrying the complementary *R* gene. Flor (1942) demonstrated that *Avr-R* gene interactions were phenotypically epistatic over "virulence-susceptibility" gene interactions. This implies that in the presence of the complementary *R* gene, the *Avr* gene product does not provide any advantage to the pathogen since it restricts the host range of the pathogen. Yet, although *Avr* genes have been identified as avirulence determinants, their primary function is expected to be associated with virulence rather than with avirulence. Indeed, evidence is accumulating that *Avr* genes encode effector proteins that contribute to the establishment of a compatible interaction between pathogen and host, either by suppressing (basal) defense responses or by interacting with host-derived virulence targets. Thus, loss of the avirulence determinant, in order to overcome *R*-gene-mediated resistance, might decrease the virulence of the pathogen. This implies that the most effective defense strategy for plants is to target *R*-gene specificity toward *Avr* genes of which the products function to condition virulence.

FUNGAL (A)VIRULENCE GENES WITH GENOTYPE AND SPECIES SPECIFICITY

Avirulence genes have been discovered by virtue of the capacity of their encoded products to induce defense responses in plants carrying the corresponding resistance gene. *Avr* genes are important determinants in the interaction between pathogen and host, as they govern host specificity. In fungus-plant interactions, 15 *Avr* genes have thus far been cloned and demonstrated to govern either genotype or species specificity (Table 4.1)

TABLE 4.1. Cloned fungal and oomycetous avirulence genes

Pathogen	Avr gene	Plant genotype carrying matching *R* gene or resistant species	AVR homology	References
Cladosporium fulvum	*Avr9*	I*Cf-9* tomato	Carboxypeptidase inhibitor[a]	Van den Ackerveken et al., 1992
	Avr4	*Cf-4* tomato	Chitin-binding protein[b]	Joosten et al., 1994; Van den Burg et al. (unpublished data)
	Avr4E	*Hcr9-4E* tomato	None	Westerink et al. (unpublished data)
	Avr2	*Cf-2* tomato	None	Luderer, Takken, et al., 2002
	Ecp1	*Cf-ECP1* tomato	Tumor necrosis factor receptor[a]	Van den Ackerveken et al., 1993; Laugé et al., 1997
	Ecp2	*CF-ECP2* tomato	None	Van den Ackerveken et al., 1993; Laugé et al., 1997
	Ecp4, 5	*Cf-ECP4, 5* tomato	None	Laugé et al., 2000
Magnaporthe grisea	*AVR-Pita*	*Pi-ta* rice	Metalloprotease (sequence motif)	Orbach et al., 2000
	PWL1, 2	Weeping love grass	None	Sweigard et al., 1995
	AVR1-CO39	*CO39* rice	None	Farman et al., 2002
Rhynchosporium secalis	*NIP1*	*Rrs1* barley	Hydrophobin[a]	Rohe et al., 1995
Phytophthora parasitica	*parA1*	*Nicotiana tabacum*	None	Ricci et al., 1992
Phytophthora infestans	*inf1*	*Nicotiana* spp.	None	Kamoun et al., 1998

[a] Structural homology, but so far no functional homology
[b] Structural homology and functional homology

The Avr *and* Ecp *Genes of* Cladosporium fulvum

Cladosporium fulvum is a biotrophic fungus that causes leaf mold of tomato (*Lycopersicon esculentum* Mill.) plants. *Cladosporium fulvum* penetrates tomato leaves through stomata and obtains nutrients via enlarged intercellular hyphae that are in close contact with the host cells (Figure 4.2). During infection, no specialized feeding

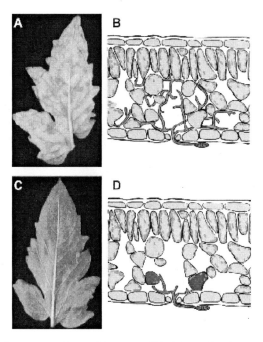

FIGURE 4.2. The compatible and incompatible interaction between tomato and *Cladosporium fulvum.* (A) Lower side of a leaf of a susceptible tomato plant that is colonized by a virulent strain of *C. fulvum* (compatible interaction). Photograph was taken two weeks after inoculation. (B) Schematic representation of a cross-section of a susceptible tomato leaf colonized by a virulent strain of *C. fulvum.* The "runner" hypha has entered the leaf through an open stoma. During colonization of the leaf, the mycelium remains confined to the extracellular space around tomato mesophyll cells. (C) Lower side of a leaf of a resistant tomato plant, two weeks after inoculation with an avirulent strain of *C. fulvum* (incompatible interaction). (D) Schematic representation of a cross-section of a resistant tomato leaf after inoculation with an avirulent strain of *C. fulvum.* The fungus is recognized as soon as a hypha enters a stoma. Recognition results in a hypersensitive response (indicated as dark cells) that restricts further growth of the fungus. (*Source:* Reprinted from Van der Hoorn, 2001, p. 11).

structures, such as haustoria, are formed. A few weeks after penetration, when intercellular spaces are fully colonized, conidiophores emerge through stomata and numerous conidia are produced that can repeat infection of healthy tomato plants. During colonization, different proteins are secreted by *C. fulvum* into the intercellular spaces between the tomato mesophyll cells. Analysis of the proteins present in the apoplast of colonized tomato leaves led to the cloning of seven genes of *C. fulvum,* all of which encode elicitor proteins. Moreover, the gene encoding elicitor protein AVR2 was cloned by a functional screening of a cDNA library of *C. fulvum* grown in vitro under starvation conditions. Four elicitor proteins (AVR2, AVR4, AVR4E, and AVR9) are race specific and trigger HR-associated defense responses in tomato plants that carry the matching *Cf* resistance gene (Joosten and De Wit, 1999). The other four elicitors, extracellular proteins ECP1, ECP2, ECP4, and ECP5, as well as ECP3, for which the encoding gene has not yet been identified, are secreted by all strains of *C. fulvum* that have been analyzed to date (Joosten and De Wit, 1999). Individual accessions within the *Lycopersicon* genus have been identified in which these ECP proteins trigger a specific HR (Laugé et al., 2000). The matching *R* genes, designated *Cf-ECPs*, present in these resistant individuals have not yet been introduced into commercial cultivars.

Race-Specific Avirulence Gene Avr2 of C. fulvum

Avirulence gene *Avr2* confers avirulence of *C. fulvum* on tomato plants carrying the *Cf-2* resistance gene. The *Avr2* gene was cloned based on HR induction of the encoded AVR2 protein in *Cf-2* tomato by functional screening of a cDNA library of *C. fulvum* grown in vitro under starvation conditions that was constructed in a binary potato virus X (PVX)-based expression vector (Takken et al., 2000) (Figure 4.3). *Avr2* encodes a cysteine-rich protein of 78 amino acids, with a predicted signal peptide of 20 amino acids for extracellular targeting (Luderer, Takken, et al., 2002). Strains of *C. fulvum* that are virulent on *Cf-2* tomato plants carry different modifications in the open reading frame (ORF) of *Avr2* (Table 4.2). In addition to a variety of different single base pair deletions or insertions, all of which result in the production of truncated AVR2 proteins, one of the modifications involves a retrotransposon insertion in the *Avr2* ORF (Luderer, Takken,

PVX::-   PVX::*Avr2*

FIGURE 4.3. Phenotype of *Cf-2* tomato plants inoculated with PVX::*Avr2* (right panel) or PVX::- (left panel). Plants were photographed 14 days after inoculation with PVX constructs. Note the AVR2-induced systemic necrosis (right panel). (*Source:* Reprinted from Luderer, R., Takken, F.L.W., De Wit, P.J.G.M., and Joosten, M.H.A.J. [2002], *Cladosporium fulvum* overcomes *Cf2*-mediated resistance by producing truncated AVR2 elicitor proteins, *Molecular Microbiology* 45: 875-884. Used with permission of Blackwell Publishing Ltd.)

et al., 2002). *Cf-2*-mediated resistance has been reported to require the *Rcr3* gene (Dixon et al., 2000). *Rcr3* was isolated by positional cloning and encodes a cysteine protease that is secreted into the apoplastic space of tomato (Krüger et al., 2002). *Rcr3* was originally identified in ethyl methyl sulfonate (EMS)-mutagenized *Cf-2* plants that either showed a partial loss *(rcr3-1)* or a complete loss *(rcr3-3)* of *Cf-2*-mediated resistance (Dixon et al., 2000). PVX-mediated expression of *Avr2* in *rcr3-1* and *rcr3-3* mutant *Cf-2* plants resulted in impaired and abolished systemic HR symptoms, respectively, suggesting a role of the extracellular Rcr3 protein in perception of AVR2 by *Cf-2* plants (Luderer, Takken, et al., 2002). Thus far, no differences have been observed between the virulence of *C. fulvum* strains lacking a functional copy of *Avr2* and similar strains that are comple-

TABLE 4.2. Overview of mutation identifed in the open reading frames of *Avr2*, *Avr4*, *Avr4E*, and *Avr9* of *Cladosporium fulvum* strains that are virulent on tomato plants that carry *Cf-2*, *Cf-4*, *Hcr9-4E*, or *Cf-9*, respectively

Mutations in *Avr2*	Codon position in ORF	Mutation in AVR2	Predicted protein*
Wild-type	–	No mutation	
C to T (stop)	66	Frame shift/stop	
D T	72	Frame shift	
D C	24	Frame shift/stop	
D A	23	Frame shift/stop	
+ A	40	Frame shift/stop	
+ A	23	Frame shift/stop	
+ Transposon	19	Insertion of 5 kB	

Mutations in *Avr4*	Codon position in ORF	Mutation in AVR4	Predicted protein
Wild-type	–	No mutation	
D C	42	Frameshift	
G to T	64	Cys-64-Tyr	
C to T	66	Thr-66-Ile	
T to C	67	Tyr-67-His	
G to T	70	Cys-70-Tyr	
G to T	109	Cys-109-Tyr	

Mutations in *Avr4E*	Codon position in ORF	Mutation in AVR4E	Predicted protein
Wild-type	–	No mutation	
T to C; T to C	82 and 93	Phe-82-Leu; Met-93-Thr	

Mutations in *Avr9*	Predicted protein
Wild-type	No mutation
Deletion of ORF	No protein No protein

* The speckled areas in the horizontal bars represent the signal peptide of the AVR proteins; open areas represent the mature part of the AVR protein; hatched areas represent the amino acid sequence encoded that follows the frameshift mutation in *Avr2* and *Avr4*. The cysteine residues are indicated as vertical lines and amino acid substitutions as dotted vertical lines. The black areas represent the amino acid sequence that is removed by N- and C-terminal processing.

mented with a functional genomic clone of *Avr2* (Luderer, Takken, et al., 2002).

Race-Specific Avirulence Gene Avr4 of C. fulvum

The AVR4 elicitor protein is secreted into the apoplastic space of tomato as a proprotein of 135 amino acids (Joosten et al., 1994). N- and C-terminal processing by fungal and plant proteases results in a mature protein of 86 amino acids (Joosten et al., 1997). The AVR4 protein contains eight cysteine residues, all of which are involved in intramolecular disulfide bonds (Van den Burg, unpublished data). As opposed to *Avr9,* the *Avr4* promoter sequence does not contain nitrogen-responsive elements, indicating that *Avr4* is regulated in a different way. During pathogenesis, however, the expression profiles of both *Avr4* and *Avr9* are similar in time and space (Van den Ackerveken et al., 1994; Joosten et al., 1997). Strains of *C. fulvum* evade *Cf-4*-mediated resistance by different single point mutations in the coding region of the *Avr4* gene (Joosten et al., 1997) (Table 4.2). These modifications result in the production of either a truncated AVR4 protein or AVR4 isoforms that exhibit single amino acid exchanges, including cysteines (Joosten et al., 1997). By using PVX-mediated expression in *Cf-4* tomato, it appears that most of these amino acid exchanges resulted in AVR4 isoforms that still exhibited necrosis-inducing activity, although this was significantly reduced when compared to the AVR4 wild-type protein (Joosten et al., 1997). These studies and supplementary data have demonstrated that all of these amino acid exchanges decrease protein stability, thereby circumventing specific recognition by *Cf-4* tomato plants (Westerink and Van den Burg, unpublished results).

AVR4 shares structural homology with invertebrate chitin-binding proteins, and binding of AVR4 to chitin oligosaccharides has been demonstrated in vitro (Van den Burg, unpublished data). AVR4 also accumulates on hyphae of *C. fulvum* during growth in the apoplastic space of tomato, most likely at positions where chitin is exposed to the surface (Van den Burg et al., unpublished data). Furthermore, an AVR4-specific high-affinity binding site (HABS) of fungal origin has been identified, which appeared to be heat and proteinase K resistant (Westerink, unpublished data). Although the latter suggests a nonproteinaceous character, AVR4 also cross-links to a fungus-derived molecule with a molecular mass of approximately 75 kDa (Westerink, unpublished data), implying that AVR4 binds either to a heat- and proteinase K-resistant protein or to another fungal protein with possibly low affinity.

It appears that only in the presence of AVR4, the highly sensitive fungus *Trichoderma viride* was protected against the antifungal activities of plant chitinases (Van den Burg et al., unpublished data). The insensitivity of *C. fulvum* to plant chitinases as well as endoglucanases in vitro (Joosten et al., 1995), however, does not depend on the production of AVR4 by the fungus, suggesting that in this case other components protected the fungus against these hydrolases. Although not measurable in vitro, AVR4 might still contribute to protect *C. fulvum* against cell wall degradation during growth *in planta.*

Race-Specific Avirulence Gene Avr4E of C. fulvum

Strains of *C. fulvum* that carry the *Avr4E* gene are avirulent on tomato plants carrying *Hcr9-4E* (a **h**omologue of *Cladosporium* **r**esistance gene *Cf-9*), which is, in addition to *Cf-4* (*Hcr9-4D*), the other functional *Cf* resistance gene present at the *Cf-4* locus (Takken et al., 1999) (Figure 4.4). The *Avr4E* gene encodes a cysteine-rich protein of 101 amino acids that is secreted into the extracellular space of tomato leaves (Westerink et al., unpublished data). Although the *Cf-4* and *Hcr9-4E* resistance genes share a high degree of overall sequence similarity (Parniske et al., 1997), their matching *Avr* gene products do not share any sequence homology. Various strains of *C. fulvum* have been identified that evade both *Cf-4-* and *Hcr9-4E*-mediated resistance. For these strains, loss of the *Avr4* avirulence function was caused by a variety of different single-point mutations in the *Avr4* allele, as mentioned earlier (Joosten et al., 1997). Loss of the *Avr4E* avirulence function appeared to be based on two different molecular mechanisms. First, strains of *C. fulvum* were identified that carry an *Avr4E* allele with two-point mutations, resulting in amino acid changes Phe62Leu and Met73Thr (AVR4E[LT]) (Westerink et al., unpublished data) (Table 4.2). In contrast to the AVR4 isoforms, this elicitor-inactive AVR4E[LT] protein is as stable as the wild-type AVR4E protein. It appeared that single amino acid substitution Phe62Leu rather than

FIGURE 4.4. Necrosis-inducing activity (NIA) of AVR4 and AVR4E elicitor proteins. *Agrobacterium* cultures carrying *Avr4* or *Avr4E* were coinfiltrated into leaves of six-week-old tobacco plants in a 1:1 ratio with *Agrobacterium* cultures carrying the resistance genes *Hcr9-4D* or *Hcr9-4E*. NIA was scored three days postinfiltration (dpi) and photographs were taken at 7 dpi. Note that AVR4 and AVR4E induce a necrotic response only in the presence of the matching resistance proteins, Hcr9-4D and Hcr9-4E, respectively.

Met73Thr causes circumvention of AVR4E recognition by *Hcr9-4E* plants (Westerink et al., unpublished data). Single-point mutations in *Avr4E*, however, which render elicitor-inactive AVR4E^L and elicitor-active AVR4E^T, have not been identified in natural virulent and avirulent strains of *C. fulvum*, respectively. Although we cannot exclude a possible simultaneous evolutionary event underlying the double amino acid substitution, strains of *C. fulvum* carrying *Avr4E^{LT}* most likely were derived from yet unidentified avirulent strains carrying *Avr4E^T*.

Surprisingly, all other strains virulent on *Hcr9-4E*-containing plants carry an *Avr4E* allele that is identical to the *Avr4E* allele in avirulent strains. It appears, however, that these strains do not secrete the AVR4E protein upon colonization of the apoplastic space of tomato. Complementation with a genomic *Avr4E* sequence of an avirulent strain of *C. fulvum* conferred avirulence on *Hcr9-4E* plants, suggesting that in this case abolished AVR4E expression resulted in circumvention of *Hcr9-4E*-mediated resistance. Indeed, no *Avr4E* transcripts could be detected when Northern blot analysis was performed on RNA isolated from a compatible interaction between these strains and tomato. Whether recombination events or (transposon) insertions within the promoter sequence of *Avr4E* cause abolished AVR4E expression still needs to be elucidated.

Race-Specific Avirulence Gene Avr9 of C. fulvum

Avr9, the first fungal *Avr* gene to be cloned and characterized, encodes a precursor protein of 63 amino acids that contains a 23-amino acid signal sequence (Van Kan et al., 1991). Upon secretion into the apoplast, AVR9 is further processed at the N-terminus by fungal and plant proteases into a mature protein of 28 amino acids, six of which are cysteines. The three-dimensional structure of AVR9, elucidated by ^{1}H-NMR, revealed that the protein contains three antiparallel β-strands that are interconnected by three disulfide bridges (Van den Hooven et al., 2001). The AVR9 protein, which contains a cysteine knot, is structurally most related to potato carboxypeptidase inhibitor (CPI) (Van den Hooven et al., 2001). AVR9, however, does not have amino acid residues identical to those located at known CPI-inhibitory sites, and thus far no protease-inhibiting activity could be detected. Structural analysis revealed that all six cysteine residues present in AVR9 are essential for its structure and necrosis-inducing activity (Kooman-Gersmann et al., 1997; Van den Hooven et al., 2001). In addition, residue Phe21, present in the solvent-exposed hydrophobic β-loop region, is also essential for the necrosis-inducing activity of AVR9 (Kooman-Gersmann et al., 1997). Moreover, when applied to transgenic *Cf-9* tobacco cell suspensions, AVR9 mutant peptide carrying Phe21Ala was incapable of inducing medium alkalization, whereas its capacity to induce an oxidative burst was reduced

(De Jong et al., 2000). Virulence of *C. fulvum* strains on *Cf-9* plants appeared to be the result of a deletion of the entire *Avr9* gene (Table 4.2). Moreover, disruption of *Avr9* by homologous recombination in *C. fulvum* strains normally avirulent on *Cf-9* plants did not affect in vitro growth or virulence of the fungus on susceptible tomato plants, suggesting that *Avr9* is dispensable for full virulence (Marmeisse et al., 1993). Although dispensable, the expression of *Avr9* is induced under nitrogen-limiting conditions in vitro (Van den Ackerveken et al., 1994; Pérez-García et al., 2001), which suggests that AVR9 might be involved in the nitrogen metabolism of the fungus. Pérez-García and colleagues (2001) identified a gene in *C. fulvum,* designated *Nrf1,* which has a strong similarity to nitrogen regulatory proteins of *Aspergillus nidulans.* Although *Nrf1*-deficient strains do not express *Avr9* under nitrogen starvation conditions in vitro, these strains are still avirulent on *Cf-9* tomato plants, suggesting that NRF1 is a major, yet not the only, positive regulator of *Avr9* expression (Pérez-García et al., 2001).

Non-Race-Specific Extracellular Protein (Ecp) *Genes of* C. fulvum

Four genes encoding extracellular proteins ECP1, ECP2, ECP4, and ECP5 have been cloned (Van den Ackerveken et al., 1993; Laugé et al., 2000). These *Ecp* genes all encode cysteine-rich proteins that are abundantly secreted by all strains of *C. fulvum* during colonization of tomato leaves. These proteins share neither sequence homology with each other nor with any sequences present in the database. Although the even number of cysteine residues present in the ECPs suggests that these residues contribute to protein stability, some (as demonstrated for ECP1 and ECP2) appear not to be involved in intramolecular disulfide bonds (Luderer, De Kock, et al., 2002). As was found for *Avr4* and *Avr9*, transcription of both *Ecp1* and *Ecp2* was strongly induced *in planta* (Wubben et al., 1994), indicating that plant-derived signals are required for the induction of both *Avr* and *Ecp* gene expression. Tomato accessions that develop a HR upon inoculation with recombinant PVX expressing *Ecp2* have been identified (Laugé et al., 1998). The responding accessions all carry a single dominant gene, designated *Cf-ECP2* gene, and show HR-associated resistance toward ECP2-producing strains of *C. fulvum* (Laugé et al., 1998). ECP1, ECP3, ECP4, and ECP5 have also been shown to act as elicitors of HR on tomato accessions and wild *Lycopersicon* plants that are resistant toward *C. fulvum,* most likely through recognition of the corresponding secreted ECP (Laugé et al., 2000). As opposed to the *Avr* genes, no modifications have thus far been found in the *Ecp* genes of naturally occurring strains of *C. fulvum*. This might be due to a lack of selection pressure on the pathogen to overcome *Cf-ECP*-mediated resistance, as the *Cf-ECP*s have not yet been introduced in commercial cultivars. On the other hand, since all strains of *C. fulvum* analyzed so far secrete the ECPs, disruption or modification of the encoding genes is thought to cause reduced virulence of the fungus on tomato. Indeed, the *Ecp2* gene appears to be required for colonization and sporulation of *C. fulvum* on mature tomato plants (Laugé et al., 1997). Moreover, *Ecp1*-deficient strains failed to sporulate as abundantly as the wild-type strain on mature tomato plants (Laugé et al., 1997). This implies that ECP1 and ECP2 are both required for full virulence of *C. fulvum* on tomato. In addition, both *Ecp1*- and *Ecp2*-deficient strains induce plant defense-associated responses more quickly and to higher levels than wild-type strains, suggesting that both ECPs are involved in the suppression of host defense-associated responses during colonization (Laugé et al., 1997). Based on this observation, an interesting parallel can be drawn with mammalian systems, in which viruses have been reported to produce extracellular suppressors of host defense responses (Laugé et al., 1997). Interestingly, ECP1 shares structural homology (based on the spacing of the cysteine residues) with a viral T2 suppressor protein as well as with the family of tumor-necrosis factor receptors (TNFRs) (Laugé et al., 1997). The T2 suppressor protein compromises the establishment of host defense responses by interacting with mediators of the immune system (tumor necrosis factors), thereby preventing its binding to endogenous TNFRs. Furthermore, a putative receptor-like kinase has been identified in plants that shares structural homology with the TNFR family. One possibility could be that ECP1 competitively inhibits the binding of defense signaling molecules to this plant receptor protein, thereby suppressing the induction of host defense responses.

The Avr *Genes of* Magnaporthe grisea

The filamentous ascomycete *Magnaporthe grisea* is the causal agent of blast disease on many species of the grass family, such as rice. *Magnaporthe grisea* initiates infection by a germinating conidium that quickly differentiates into a specialized cell, the appressorium. Once mature, the melanized appressorium generates enormous hydrostatic pressure which forces a narrow penetration peg through the plant cuticle and epidermal cell wall. After penetration, the fungus grows intracellularly and produces sporulating lesions within five to seven days.

Genotype-Specific Avr *Genes of* M. grisea

Strains of *M. grisea* that carry the *Avr-Pita (AVR2-YAMO)* gene are avirulent on rice cultivars that carry the corresponding *R* gene *Pi-ta* (Orbach et al., 2000). The *Avr-Pita* gene is located very close to the telomere of chromosome 3 and encodes a predicted polypeptide of 223 amino acids. AVR-Pita exhibits substantial similarity to NPII, a neutral zinc metalloprotease from *Aspergillus oryzae*. Based on this homology, the N-terminus of AVR-Pita was predicted to be further processed to an active form of 176 amino acids. This AVR-Pita$_{176}$ protein, but not the intact AVR-Pita$_{223}$ protein and AVR-Pita$_{166}$ (which has an additional deletion at the N-terminus), triggers the *Pita*-dependent HR when produced inside rice cells by transient expression (Jia et al., 2000). In the region that corresponds to the consensus zinc-binding domain of neutral zinc metalloproteases, residue Glu-177 of AVR-Pita$_{223}$ (i.e., Glu-130 of AVR-Pita$_{176}$) is predicted to be essential for metalloprotease activity. Interestingly, replacement of this Glu residue by Asp, as found in spontaneous gain of virulence mutants, abolishes the HR-inducing ability of AVR-Pita$_{176}$ (Jia et al., 2000; Orbach et al., 2000). This implies that the protease activity of *Avr-Pita*, although not yet biochemically demonstrated, plays an essential role in avirulence (Orbach et al., 2000). The majority of spontaneous virulent mutants of *M. grisea* carry deletions ranging from 100 bp up to 10 kb, which is consistent with the genetic instability observed for genes that are located at a telomere (Orbach et al., 2000). In addition to point mutations and deletions, gain of virulence on

Pi-ta rice cultivars was also mediated by an insertion of a pot3 transposon into the promoter of *Avr-Pita* (Kang et al., 2001). Despite its putative metalloprotease activity, no role in virulence could yet be assigned to AVR-Pita.

The *AVR1-CO39* gene of *M. grisea* has been identified as the minimal (1.05 kb) fragment that confers avirulence on rice cultivar CO39 (Farman et al., 2002). Only a small number of rice-infecting *M. grisae* isolates from the Philippines, however, are avirulent on this cultivar. Although most virulent isolates lack the entire *AVR1-CO39* locus, it appeared that in some cases complex genomic rearrangements have occurred at the *AVR1-CO39* locus, each of which resulted in nonfunctional alleles (Farman et al., 2002).

Species Specificity Conferred by PWL *Genes of* M. grisea

The *PWL2* (for **p**athogenicity toward **w**eeping **l**ovegrass) gene of *M. grisea* determines host-species specificity. Strains of the fungus expressing *PWL2* are avirulent on weeping lovegrass but virulent on rice and barley (Sweigard et al., 1995). *PWL2* encodes a glycine-rich protein of 145 amino acids with a putative signal peptide for extracellular targeting. Analysis of spontaneous virulent mutants on weeping lovegrass revealed that the *PWL2* allele is genetically unstable, although it is not located at a telomere (Sweigard et al., 1995). As found for the *Avr-pita*-deficient mutants, spontaneous deletion of *PWL2* had no apparent effect on virulence under laboratory conditions. Strains of *M. grisea* also evade PWL2 recognition by a single base pair change that results in the creation of a putative N-glycosylation site. This PWL2 mutant protein exhibited reduced elicitor activity either due to glycosylation or due to the amino acid change itself.

The *PWL2* gene is a member of a rapidly evolving gene family of which the homologue *PWL1* and the allelic *PWL3/PWL4* genes map at different chromosomal locations (Kang et al., 1995). The PWL2 protein is 75 percent identical to the PWL1 protein, and 51 and 57 percent identical to the PWL3 and PWL4 proteins, respectively. As opposed to *PWL1* and *PWL2,* the *PWL3* and *PWL4* genes are nonfunctional *Avr* genes, since they do not confer avirulence on weeping lovegrass. In contrast to *PWL3*, *PWL4* becomes functional in preventing infection of weeping lovegrass when its expression is driven by either the *PWL1* or the *PWL2* promoter (Kang et al., 1995). This indicates that *PWL4* encodes a functional AVR protein, which is not recognized by weeping lovegrass due to lack of expression of the gene.

The Avr *Genes of* Rhynchosporium secalis

The fungus *Rhynchosporium secalis* is known as the causal agent of leaf scald on barley, rye, and other grasses. *Rhynchosporium secalis* initiates infection by penetrating the cuticle, followed by extracellular growth of hyphae between the cuticle and the outer epidermal cell walls. The fungus develops an extensive subcuticular stroma, causes an early collapse of a few epidermal cells and the underlying mesophyll cells, and finally starts to sporulate. Among the secreted proteins in culture filtrates of *R. secalis,* a class of necrosis-inducing proteins (NIPs) has been identified that induces necrosis in certain barley cultivars and other cereals (Wevelsiep et al., 1993). The phytotoxicity of these NIPs, which is associated with lesion development, appeared to be based on their stimulatory effect on the plant plasmalemma H^+-ATPase, probably in order to release plant nutrients (Wevelsiep et al., 1993).

Strains of *R. secalis* that secrete NIP1 are unable to grow on barley cultivars that carry the *Rrs1* gene. *Rrs1*-mediated resistance is not associated with a rapid HR but with the accumulation of mRNAs encoding peroxidase and PR proteins of the PR-5 class (Rohe et al., 1995). The *NIP1* gene encodes a secreted elicitor-active protein of 60 amino acids, ten of which are cysteine residues (Rohe et al., 1995). The three-dimensional structure of NIP1 has revealed that all ten cysteines form intramolecular disulfide bonds, providing stability to the protein (Vant' Slot, unpublished data). The spacing pattern of the first eight cysteine residues in NIP1 (-C-CC-C-C-CC-C-) has also been found in another class of fungal proteins, the hydrophobins. The partially resolved disulfide bond pattern of the *Ophiostoma ulmi* hydrophobin, however, differs from that of NIP1, suggesting that NIP1 is not functionally related to the hydrophobins. Thus far, no structural homology has been found between NIP1 and other proteins (Vant' Slot, unpublished data). Several avirulent races of *R. secalis* carry three amino acid changes in the *NIP1* gene product (NIP1 type II).

Despite the fact that elicitor activity is reduced, these NIP1 type II proteins still confer avirulence and still exhibit toxicity. Two elicitor-inactive NIP1 proteins (NIP1 type III and IV proteins) have been identified to carry two different, additional amino acid substitutions (Rohe et al., 1995). Moreover, virulence of *R. secalis* on *Rrs1* barley plants was accomplished by deletion of the entire *NIP1* gene (Rohe et al., 1995). These strains lacking *NIP1* were less virulent on susceptible barley cultivars than those carrying *NIP1*, demonstrating that NIP1 plays a role in virulence of *R. secalis* and that toxic activity of NIP1 type III and IV is retained (Rohe et al., 1995).

The Genes Encoding Elicitins of Phytophthora *spp.*

Oomycetous plant pathogens, such as *Phytophthora* spp., downy mildews, and *Pythium* spp., cause devastating diseases on numerous crops and ornamental plants. In the middle of the nineteenth century, *Phytophthora infestans* destroyed potato crops in Ireland, which resulted in starvation and decimation of the Irish population. Despite the fact that many *R* genes have been incorporated into potatoes through traditional breeding strategies, the late-blight pathogen has remained a continuous threat for potato growers worldwide because of its adaptive abilities to overcome these genes.

Although oomycetes exhibit filamentous growth, they share little taxonomic affinity to filamentous fungi and are more closely related to eukaryotic algae (Kamoun, Huitema, and Vleeshouwers, 1999). The disease cycle of oomycetes starts when zoospores encyst and germinate on root or leaf surfaces. In some species, sporangia germinate directly. Germ tubes penetrate the epidermal cell layer, secondary hyphae expand through the intercellular space to neighboring cells, and in some cases feeding structures are formed inside the mesophyll cells. The major defense reaction in resistant plants to many *Phytophthora* and downy mildew species is associated with a HR. Partial resistance to *Pythium,* however, appeared to be mediated by physical barriers rather than by a HR (Kamoun, Huitema, and Vleeshouwers, 1999).

Phytophthora infestans, as well as other *Phytophthora* and *Pythium* species, produce extracellular proteins of 10 kDa, termed elicitins, which contain three highly conserved disulfide bridges (Huet et al.,

1995; Boissy et al., 1996). It has been demonstrated that elicitins bind to sterols and mediate their transfer between micelles and artificial phospholipid membranes (Mikes et al., 1998). Because *Phytophthora* species do not synthesize sterols themselves, elicitins might contribute to the assimilation and growth of the oomycete. Elicitins induce non-genotype-specific defense-associated responses, including a HR, in plants of the genus *Nicotiana* (i.e., Solanaceae) and in some cultivars of radish, turnip, and rape (i.e., Cruciferae).

Species-Specific Gene parA1 *of* Phytophthora parasitica

In *Phytophthora parasitica,* the absence of elicitin production was correlated with high virulence on tobacco. Although elicitins are encoded by a multigene family, it appeared that *parA1* was the main elicitin-encoding gene expressed in vitro and *in planta* by *P. parasitica*. The *parA1* gene was cloned from *P. parasitica* and encoded a secreted protein (parasiticein) of 98 amino acids, of which six were cysteine residues (Ricci et al., 1992). The *parA1* gene has been proposed to act as a species-specific *Avr* gene, since it triggers HR-mediated resistance toward *P. parasitica* in *Nicotiana tabacum* (tobacco). Elicitin-producing *P. parasitica* isolates have been demonstrated to cause disease on tobacco, but not on tomato, upon down-regulation of parA1 expression (Colas et al., 2001). It appears that this down-regulation event relies on a mechanism that is dependent on the *P. parasitica* genotype rather than the host plant (Colas et al., 2001).

Genotype-Specific Gene inf1 *of* Phytophthora infestans

The *inf1* gene of *P. infestans* encoding infestin was cloned by screening a cDNA library of a compatible interaction between *P. infestans* and potato with a *parA1* gene fragment (Kamoun et al., 1997). While high levels of *inf1* transcripts were observed in mycelium grown in vitro, the expression of *inf1* was down-regulated *in planta* (Kamoun et al., 1997). As opposed to *parA1,* down-regulation of *inf1* expression is not required to evade plant defense responses, since the host plant potato does not respond to INF1 protein (Kamoun et al., 1997). In leaves of nonhost *Nicotiana* plants, however, injection of INF1 induces a specific HR. To determine whether INF1 plays a role in nonhost resistance, members of the *Nicotiana* family were inoculated with *inf1*-deficient *P. infestans* strains. These *inf1*-deficient strains were able to cause disease when inoculated on leaves of *N. benthamiana,* whereas on other *Nicotiana* species, such as *N. tabacum* (tobacco), these *inf1*-deficient strains were still avirulent (Kamoun et al., 1998). This demonstrated that INF1 confers nonhost resistance toward *P. infestans* in *N. benthamiana,* but was not the main determinant of nonhost resistance to *P. infestans* in other *Nicotiana* species. Putative additional candidates that confer avirulence on to-bacco are the products of the *inf2A* and *inf2B* genes of *P. infestans,* which also induced a HR when injected into *N. tabacum* leaves (Kamoun et al., 1998).

PATHOGENS WITH EVOLVED MECHANISMS TO COUNTERACT PLANT DEFENSE RESPONSES

As described previously, plant pathogens evade *R*-gene-mediated resistance by modification of the elicitor proteins either by mutations in, or deletion of, the *Avr* genes or by (down-)regulation of *Avr* gene expression. Yet, when the circumvention of elicitor detection fails, or when the elicitor component is essential for virulence, pathogens require mechanisms to subvert the induced plant defense responses. Indeed, plant pathogens counteract plant defenses by secreting enzymes that detoxify defense compounds, including phytoalexins, or use ATP-binding cassette (ABC)-transporters to mediate the efflux of toxic compounds (as reviewed by Idnurm and Howlett, 2001). Moreover, some bacterial pathogens interfere with *R*-gene-mediated resistance by secreting proteins that "mask" the presence of a particular AVR effector protein (Ritter and Dangl, 1996).

Recent studies have revealed that some pathogenic fungi have evolved counter-defense mechanisms that enable the suppression of plant defense responses. This mechanism involves a class of proteins, termed glucanase inhibitor proteins (GIPs), which are secreted by *Phytophthora sojae* f. sp. *glycines* and which inhibit the endoglu-canase (EnGL) activity of its soybean host (Rose et al., 2002). Sequences homologous to GIPs have been identified in genomic DNA of other *Phytophthora* species, whereas several other plant pathogenic fungi do not exhibit related sequences (Rose et al., 2002). GIPs are homologous to the trypsin class of serine proteases but do not exhibit proteolytic activity. The basis of endoglucanase inhibition by GIPs involves the formation of a stable complex. This association appears to be specific, since GIP1 inhibited soybean endoglucanase-A (EnGL-A) but not EnGL-B. GIPs also suppressed the release of elicitor-active oligoglucosides from *P. sojae* mycelial walls in vitro and during pathogenesis in vivo (Rose et al., 2002). Thus GIPs suppress cell wall disassembly by inhibiting endoglucanases and suppress the activation of defense-associated responses by inhibiting the release of oligoglucoside elicitors (Rose et al., 2002).

The *CgDN3* gene of the hemi-biotrophic pathogen *Colletotrichum gloeosporioides,* which encodes an extracellular protein of 54 amino acids, is induced under nitrogen starvation conditions at the early stage of infection on tropical pasture legume (Stephenson et al., 2000). On intact leaves of a normally compatible host, *CgDN3*-disrupted mutants were nonpathogenic and elicited a hypersensitive-like response, the latter of which may be part of a basal defense reaction toward *C. gloeosporioides* (Stephenson et al., 2000). When applied to wound sites, however, these mutants were able to grow necrotropically on host leaves and form necrotic spreading lesions. It was therefore suggested that at the early stages of the primary infection process, CgDN3 is required for pathogenicity and functions to suppress the hypersensitive-like response (Stephenson et al., 2000). As opposed to the *C. fulvum* ECP1 and ECP2 elicitors that also suppress host defense responses (Laugé et al., 1997), no elicitor function has been assigned to CgDN3.

PLANT GENES THAT CONFER RESISTANCE
TOWARD FUNGI AND OOMYCETES

R genes are very abundant in plant genomes and in most cases they belong to tightly linked gene families. The *R* genes that have been characterized can be divided into six classes of proteins (reviewed by Takken and Joosten, 2000) (Table 4.3). Three of these classes contain leucine-rich repeats (LRRs), of which the class of nucleotide-binding site (NB)-LRRs proteins is the most abundant. The NB-LRR class

TABLE 4.3. R proteins identified in gene-for-gene interactions confering resistance toward fungi and oomycetes classified based on homologous structural domains

R gene	Pathosystem	Structure	Matching *Avr* gene	References
L6, M, N, P	Flax/*Melampsora lini*	TIR-NB-LRR	U/N[a]	Islam and Mayo, 1990
RPP1, 10, 14	Arabidopsis/*Peronospora parasitica*	TIR-NB-LRR	U/N	Botella et al., 1998
RPP4, 5	Arabidopsis/*Peronospora parasitica*	TIR-NB-LRR	U/N	Van der Biezen et al., 2002
RPP8	Arabidopsis/*Peronospora parasitica*	CC-NB-LRR	U/N	McDowell et al., 1998
RPP13	Arabidopsis/*Peronospora parasitica*	CC-NB-LRR	U/N	Bittner-Eddy et al., 2000
Mla	Barley/*Erysiphe graminis*	CC-NB-LRR	U/N	Halterman et al., 2001
R1	Potato/*Phytophthora infestans*	CC-NB-LRR	U/N	Ballvora et al., 2002
Dm3	Lettuce/*Bremia lactucae*	NB-LRR	U/N	Anderson et al., 1996
I2	Tomato/*Fusarium oxysporum*	NB-LRR	U/N	Ori et al., 1997
Rp1	Maize/*Puccinia sorghi*	NB-LRR	U/N	Richter et al., 1995
Pi-ta	Rice/*Magnaporthe grisea*	NB-LRD	*Avr-Pita*	Bryan et al., 2000
Cf-2	Tomato/*Cladosporium fulvum*	LRR-TM	*Avr2*	Dixon et al., 1996
Cf-4	Tomato/*Cladosporium fulvum*	LRR-TM	*Avr4*	Thomas et al.,1997
Hcr9-4E	Tomato/*Cladosporium fulvum*	LRR-TM	*Avr4E*	Takken et al., 1999
Cf-5	Tomato/*Cladosporium fulvum*	LRR-TM	U/N	Dixon et al., 1998

TABLE 4.3. *(continued)*

R gene	Pathosystem	Structure	Matching *Avr* gene	References
Cf-9	Tomato/*Cladosporium fulvum*	LRR-TM	*Avr9*	Jones et al.,1994
Rpw8	Arabidopsis/*Erysiphe Cichoracearum*	CC-TM	U/N	Xiao, Ellwood, et al., 2001
Ve1, Ve2	Tomato/*Verticillium dahliae*	CC-LRR-TM-PEST	U/N	Kawchuk et al., 2001

a U/N = unknown

TIR, Toll/interleukin 1 receptor-like domain; NB, nucleotide binding site; LRR/LRD, leucine-rich repeat/domain; CC, coiled-coil; TM, transmembrane domain; and PEST, Pro-Glu-Ser-Thr sequences

can be further subdivided, based on the deduced N-terminal features of the R protein. The N-terminus of one subclass contains a TIR domain, which has homology to the *Drosophila Toll* and mammalian *Interleukin-1* receptors, while the other contains putative coiled-coil (CC) domains. The TIR-NB-LRR and CC-NB-LRR proteins confer resistance to a broad range of pathogens, including viruses, bacteria, fungi, oomycetes, nematodes, and insects.

The other two classes of LRR proteins involve the LRR-trans-membrane-anchored (LRR-TM) proteins and the LRR-TM-kinase proteins. The LRR-TM proteins have been demonstrated to provide ·resistance toward fungi and nematodes, whereas the LRR-TM-kinase (i.e., Xa-21) confers bacterial (*Xanthomonas oryzae* pv. *oryzae*) resistance.

Members of the fourth class of *R* genes represent protein kinases which, in the case of *Pto,* confer resistance toward *Pseudomonas syringae* pv. *tomato* carrying *AvrPto*. To date, *R* genes against fungi that encode protein kinases have not been identified, yet their involvement in defense-signaling pathways leading to fungal disease resistance cannot be excluded.

The fifth class of *R* genes is represented by *RPW8,* a small, putative membrane protein with a possible cytoplasmic coiled-coil domain that confers powdery mildew (*Erysiphe cichoracearum*) resistance in *Arabidopsis* (Xiao, Ellwood, et al., 2001).

The tomato verticillium wilt *Ve* resistance genes from tomato represent the sixth class of *R* genes (Kawchuk et al., 2001). The two closely linked *Ve1* and *Ve2* genes, whose products might recognize different ligands, encode TM-surface glycoproteins having an extracellular LRR domain, endocytosis-like signals, and leucine zipper (LZ) or Pro-Glu-Ser-Thr (PEST) sequences. The LZ can facilitate dimerization of proteins through the formation of CC structures, while PEST sequences are often involved in ubiquitination, internalization, and degradation of proteins. Receptor-mediated endocytosis could provide a mechanism through which cells capture ligands and remove signaling receptors from the cell surfaces (Kawchuk et al., 2001).

The NB-LRR Class of R Proteins

In flax, NB-LRR proteins have been identified that mediate recognition of 31 different rust *(Melampsora lini)* strains. These different resistance specificities are distributed among five polymorphic loci: *K, L, M, N,* and *P* (Islam and Mayo, 1990). Comparative sequence analysis of the R proteins encoded by the flax *L* alleles has demonstrated that both LRR and TIR domains play a role in determining resistance specificities (Luck et al., 2000). Flax *R* genes confer resistance only to those strains of *M. lini* that carry the corresponding *Avr* gene; however, none of the 31 genetically defined *Avr* genes has been cloned yet.

In *Arabidopsis*, several *R* gene loci have been identified that confer resistance toward strains of the oomycetous pathogen *Peronospora parasitica*. These *RPP* loci comprise genes that encode TIR-NB-LRR proteins (RPP1, 4, 5, 10, and 14) and CC-NB-LRR proteins (RPP8 and RPP13) (Botella et al., 1998; McDowell et al. 1998; Bittner-Eddy et al., 2000; Van der Biezen et al., 2002). The *RPP4* gene of the *Arabidopsis* landrace Columbia (Col) is an orthologue of the *RPP5* gene of Landsberg *erecta* (L*er*). *RPP4* confers resistance to *P. parasitica* races Emoy2 and Emwa1, whereas *RPP5* confers resistance toward these races as well as toward *P. parasitica* race Noco2 (Van der Biezen et al., 2002). These strains of *P. parasitica* might carry two distinct, as yet not identified, *Avr* determinants, i.e., *AvrRPP4* and

AvrRPP5 that are recognized by RPP4 and RPP5, respectively. *RPP4* and *RPP5*, on the other hand, could also have overlapping specificities, whereby AvrRPP4, but not AvrRPP5, is recognized by both RPP4 and RPP5.

Other *R* genes that belong to the NB-LRR class, which are also members of complex resistance loci, confer multiple resistance specificities toward fungal pathogens as well. In barley, the *Mla* locus confers resistance against various races of powdery mildew (*Blumeria graminis* f. sp. *hordei*) on a gene-for-gene basis (Halterman et al., 2001). Moreover, the *Rp1* rust *(Puccinia sorghi)* locus in maize (*Zea mays* L.) and the *Dm3* downy mildew *(Bremia lactucae)* locus in lettuce (*Lactuca sativa* L.) all contain multiple genetically linked resistance specificities (Richter et al., 1995; Anderson et al., 1996). For all of these gene-for-gene interactions, no matching *Avr* genes have thus far been characterized.

The rice blast gene *Pi-ta* confers gene-for-gene-based resistance against strains of *M. grisea* that express *AVR-Pita*. *Pi-ta* encodes a putative cytoplasmic receptor and is a member of the NB-receptor class of *R* genes. *Pi-ta* lacks an N-terminal TIR or CC domain and the C-terminal leucine-rich domain (LRD) lacks the characteristic LRR motif found in other proteins of the NB-LRR class. The predicted protein encoded by *Pi-ta* present in resistant rice varieties differs by only one amino acid, located at position 918, from the protein encoded by susceptible rice varieties (Bryan et al., 2000). Moreover, it appears that Pi-ta protein with alanine-918, but not Pi-ta with serine-918, interacts with AVR-Pita in the cytoplasm of rice cells to induce resistance responses.

The TM-LRR Class of R Proteins

Another class of *R* genes includes the *Cf* genes of tomato, which confer resistance to strains of *C. fulvum* carrying the corresponding *Avr* gene (Table 4.3). The *Cf* genes encode proteins with a predicted signal peptide for extracellular targeting, a LRR region, a TM domain, and a short cytoplasmic tail (Jones and Jones, 1996; Joosten and De Wit, 1999). The *Cf* genes are members of multigene families and have been designated *Hcr2* or *Hcr9* (for **h**omologues of *Clado-sporium* **r**esistance genes *Cf-2* and *Cf-9*, respectively). Two nearly identical *Cf-2* genes (*Cf-2.1* and *Cf-2.2*) and *Cf-5 (Hcr2-5C)* confer resistance toward *C. fulvum* strains through recognition of the *Avr2* and *Avr5* gene products, respectively (Dixon et al., 1996, 1998). The *Cf-4* and *Cf-9* gene clusters each consist of five *Hcr9* homologues, of which *Hcr9-4D* (i.e., *Cf-4*), *Hcr9-4E,* and *Hcr9-9C* (i.e., *Cf-9*) gene products specifically recognize AVR4, AVR4E, and AVR9, respectively (Jones et al., 1994; Thomas et al., 1997; Takken et al., 1999).

PERCEPTION OF AVR GENE PRODUCTS BY RESISTANT PLANT GENOTYPES

In the absence of the corresponding *R* gene, many *Avr* genes have been demonstrated to provide a selective advantage to the pathogen (Laugé and De Wit, 1998; White et al., 2000; Staskawicz et al., 2001; Bonas and Lahaye, 2002). This, together with the maintenance of *Avr* genes within pathogen populations, implies that the primary function of *Avr* genes is to confer virulence to the pathogen. Although for most *Avr* gene products no biological function has yet been defined, it appears that in colonized plant tissue, AVR proteins colocalize with various host virulence targets. Together with the fact that proper subcellular targeting is essential for the avirulence activity of AVRs, this implies that AVR proteins, virulence targets, and R proteins are possibly part of one complex (Van der Hoorn et al., 2002). Four models have been proposed to address the question of how and where AVR proteins are recognized by resistant plant genotypes (Figure 4.1). These models will be discussed in detail in the following section.

Direct Perception of AVR Proteins

This model reflects the most simple interpretation of Flor's gene-for-gene hypothesis: a classical receptor-ligand model that predicts a direct interaction between *Avr* and *R* gene products (Figure 4.1). Thus far, direct physical interaction has been demonstrated for only Pto from tomato and AvrPto from *Pseudomonas syringae* (Tang et al., 1996), and for Pi-ta from rice and AVR-Pita from *M. grisae* (Jia et al., 2000). The cytoplasmic localization of both Pto and Pi-ta is consistent with the observation that *AvrPto* and *Avr-Pita* induce a HR when expressed inside plant cells (Scofield et al., 1996; Jia et al., 2000). Pto is a Ser/Thr kinase that interacts with and phosphorylates a second Ser/Thr kinase, Pti1, and several defense-related transcription factors, such as Pti4, Pti5, and Pti6 (Martin et al., 1993; Xiao, Tang, and Zhou, 2001). The most straightforward explanation for induction of plant defense responses would be that binding of AvrPto to Pto leads to a conformational change of the kinase protein which in turn triggers downstream defense signaling pathways. Pto function, however, requires *Prf*, a NB-LRR-encoding gene (Salmeron et al., 1996). Therefore, the "coreceptor" and/or the "guard" model was put forward to rationalize the mechanism of AvrPto-induced defense activation (Figure 4.1). AVR-Pita recognition, on the other hand, is mediated by direct interaction with the LRD of Pi-ta (Bryan et al., 2000; Jia et al., 2000). It appears that the putative metalloprotease activity of AVR-Pita is required for its direct interaction with Pi-ta (Jia et al., 2000), suggesting a protease-dependent defense elicitation model (Figure 4.1). Possibly, Pi-ta contains protease cleavage sites, which upon proteolytic processing renders an active form either by a conformational change or by the release of elicitor peptide(s) that trigger(s) defense responses.

Indirect Perception of AVR Proteins

In addition to direct perception of AVR proteins by R proteins, three models have been proposed that postulate an indirect interaction between *Avr* and *R* gene products can take place. These models include enzymatic- or protease-dependent defense elicitation, or the involvement of a "coreceptor" or a "guard" protein in triggering defense responses.

Protease-Dependent Defense Elicitation Model

Rcr3, a gene required for *Cf-2*-mediated resistance toward *C. fulvum* strains carrying *Avr2*, has recently been cloned and found to encode a tomato cysteine endoprotease (Krüger et al., 2002). The fact that Rcr3 is secreted into the apoplastic space is consistent with the extracellular localization of AVR2 and Cf-2. Several protease-dependent mechanisms underlying the *Cf-2*-mediated resistance can be envisaged, whereby Rcr3 most likely functions upstream of Cf-2. Rcr3 might process AVR2 to generate a mature ligand, or Rcr3 might degrade AVR2, thereby releasing active elicitor peptides that interact with the extracellular LRR of Cf-2. This would imply that not AVR2 itself but rather a protease-dependent signal triggers *Cf-2*-mediated resistance. Another possibility is that AVR2 forms a complex with Rcr3, which subsequently triggers *Cf-2*-dependent downstream signaling pathways that lead to disease resistance. Rcr3, on the other hand, might also be part of a basal defense mechanism of the plant, which upon inhibition by AVR2, turns on *Cf-2*-mediated defense responses.

The "Coreceptor" Model

Most *R* genes encode proteins that carry LRR domains. These LRRs, located either intracellular (in the case of most NB-LRR proteins) or extracellular (in the case of LRR-TM proteins), have been implicated to function in protein-protein interactions (Jones and Jones, 1996). NB-LRR proteins often function together with Ser/Thr kinases to trigger *R*-mediated signal transduction pathways, as demonstrated for Pto kinase, which requires the NB-LRR protein Prf (Salmeron et al., 1996), as well as for the NB-LRR protein RPS5, which requires PBS1 kinase (Swiderski and Innes, 2001). Moreover, in the development of shoot apical meristem, CLV1, a receptor-like protein kinase with extracellular LRRs, and CLV2, a protein with predicted extracellular LRRs and a short cytoplasmic domain, function together to recognize an extracellular peptide encoded by *CLV3* (Rojo et al., 2002). Interestingly, the *Cf* gene family is similar to *CLV2* in that it encodes plasma membrane anchored proteins with predicted extracellular LRRs and short cytoplasmic regions which lack any obvious downstream signaling domains. Recently, it has been demonstrated that both Cf-4 and Cf-9 are part of a heteromultimeric membrane-associated complex of approximately 420 kDa (Rivas, Romeis, and Jones, 2002; Rivas, Mucyn, et al., 2002). Based on the analogy between the two systems, it may be possible that a component homologous to CLV1, i.e., a receptor-like kinase, is required for *Cf*-mediated perception of extracellular AVR proteins (Joosten and De Wit, 1999) (Figure 4.1).

The "Guard" Hypothesis

One function of AVR proteins that is likely to contribute to virulence of the pathogen is through manipulation of certain (virulence) targets present in the host. The ongoing battle between plants and pathogens could in turn have led to a development of strategies in which R proteins act as "guards" to monitor the behavior of molecules that are targets of AVR proteins. Van der Biezen and Jones (1998) have proposed that the function of AvrPto for *P. syringae* is to target Pto and suppress the nonspecific defense pathway induced by this kinase. The AvrPto-Pto complex or AvrPto-activated Pto is recognized by Prf, which subsequently initiates defense responses. A variety of mutations have been identified that disrupt the avirulence function of AvrPto without affecting its virulence function (Chang et al., 2001). Moreover, these mutants failed to interact with Pto, which, in line with the "guard" hypothesis, implies that AvrPto interacts with virulence targets other than Pto.

The "guard" hypothesis could also explain why, in spite of the fact that extracellular perception of AVR9 is consistent with the predicted extracellular location of Cf-9, no direct interaction between AVR9 and Cf-9 has been detected (Luderer et al., 2001). Moreover, an AVR9-specific high-affinity binding site has been identified in plasma membranes of susceptible *Cf-0* as well as resistant *Cf-9* tomato plants, implying the involvement of a third component in *Cf-9*-mediated perception of AVR9 (Kooman-Gersmann et al., 1996). One possibility could be that the HABS represents the virulence target of AVR9. Cf-9 may "guard" this virulence target, sense its modification by AVR9, and trigger downstream defense responses that lead to resistance (Van der Hoorn et al., 2002). Recent studies with tobacco cell suspensions have revealed that *Cf-9*-mediated defense responses are attenuated at temperatures higher than 20°C and are completely suppressed at 33°C (De Jong et al., 2002). Interestingly, a correlation was found between the temperature sensitivity of this response and the amount of AVR9-HABSs, suggesting that the HABS is unstable at elevated temperatures (De Jong et al., 2002).

The *Cf-4*-mediated defense responses are also temperature sensitive; yet, as opposed to AVR9, thus far no plant-derived AVR4-specific HABS has been detected (Westerink, unpublished results). Since AVR4

exhibits chitin-binding activity, an interaction with chitin oligosaccharides might be required for AVR4 recognition by *Cf-4* plants.
However, *Cf-4*-mediated defense responses are also triggered by
AVR4 in the absence of chitin (De Jong et al., 2002), suggesting a
perception mechanism independent of chitin. Thus, in order to be
consistent with the "guard" hypothesis, AVR4 may function to bind to
chitin, as well as to a host-encoded virulence target that is guarded by
Cf-4.

Virulence targets can also include regulatory proteins which are involved in signal transduction pathways that activate basal or specific
plant defense responses. Mutational analysis of *Arabidopsis* has
identified several genes required for *NB-LRR* gene-mediated resistance toward the oomycete *P. parasitica* as well as the bacterium
P. syringae (McDowell et al., 2000; Dangl and Jones, 2001). One of
these genes, *RIN4,* has been proposed to be a target of the virulence
activities of two elicitor proteins of *P. syringae,* AvrRpm1 and AvrB
(Mackey et al., 2002). In the absence of RIN4, plants exhibited enhanced resistance toward *P. syringae* and *P. parasitica,* indicating
that RIN4 negatively regulated the basal defense-signaling pathway.
It has been suggested that interaction with and/or phosphorylation of
RIN4 by AVR proteins enhances the activity of RIN4 as a negative
regulator of basal plant defense. Thus, in the absence of RPM1, the
AVR proteins function as virulence factors by manipulating RIN4
and suppressing basal defense mechanisms, whereas in resistant
plants, manipulation of RIN4 by AVR proteins is sensed by RPM1,
which subsequently mounts a HR (Mackey et al., 2002).

The *RPP4* and *RPP5* resistance genes of *Arabidopsis thaliana*
confer resistance to *P. parasitica.* The encoding proteins both interact
with AtRSH1, a predicted cytoplasmic molecule with significant
homology to bacterial RelA and SpoT proteins (Van der Biezen et al.,
2000). These RelA/SpoT proteins function as rapidly activated transcription cofactors in bacteria. AtRSH1 is proposed to mediate transcriptional activation of stress- and defense-related genes and compounds (Van der Biezen et al., 2000). In line with the "guard"
hypothesis, pathogen-derived AVR proteins might interfere with the
function of AtRSH1, and RPP5 might have evolved to specifically
recognize this physical association and subsequently activate defense
responses (Van der Biezen et al., 2000).

The elicitin proteins of *Phytophthora* spp. behave similar to sterol carrier proteins. They can bind and pick up sterols from plasma membranes (Mikes et al., 1998). The ability of elicitins to load and transfer sterols correlates with their HR-inducing elicitor activity (Osman et al., 2001). Moreover, mutations that affect the affinity of different elicitins for sterol also seem to affect the affinity for the HABS identified on plasmamembranes of tobacco cells (Osman et al., 2001). A model has been proposed in which binding of elicitins to the HABS requires the formation of a sterol-elicitin complex (Osman et al., 2001). Thus, the HABS, which is composed of two plasmamembrane *N*-glycoproteins of 50 kDa and 162 kDa (Bourque et al., 1999), most likely evolved to recognize the sterol-elicitin complex. This heterodimeric HABS is also detected in plasma membranes of *Arabidopsis,* which is nonresponsive toward elicitins (Bourque et al., 1999), indicating that additional components are required to trigger defense responses in *Arabidopsis.* One possibility could be that a functional, yet unidentified, R protein is present in tobacco, but not in *Arabidopsis,* which "guards" the interaction of the sterol-elicitin complex with the heterodimeric HABS and thereby confers disease resistance.

DEVELOPMENT OF TRANSGENIC PLANTS THAT DISPLAY BROAD RESISTANCE AGAINST PATHOGENS

Genetic analysis of plant disease resistance has demonstrated that single *R* gene products control resistance by mediating specific recognition, either directly or indirectly, of complementary AVR proteins produced by pathogens. Plant breeders have often used *R* genes to introduce resistance into cultivated crops. However, the introgressed *R* genes, with only a few exceptions, have been shown to lack long-term durability in the field (Stuiver and Custers, 2001). Early evidence arose from studies on the rust fungi *P. graminis* f. sp. *tritici* and *M. lini* which suggested a possible relationship between the durability of *R* genes and pathogen variation; *Avr* genes with low mutation rates corresponded to more durable resistance (Flor, 1958). Flor (1958) proposed that easily mutated *Avr* genes in pathogen populations were likely to be less critical to pathogen fitness than those that mutate rarely. Thus, if loss of avirulence function is associated with a reduced virulence of the pathogen, the complementary R gene is durable. Therefore, in order to confer durable resistance in crop plants, there is a considerable interest in cloning R genes, such as *Cf-ECP2* (Laugé et al., 1998), which target *Avr* genes that are important for virulence.

De Wit (1992) proposed a strategy, referred to as the two-component sensor system, to apply *Avr* genes in molecular resistance breeding. Thereby, an *Avr* gene is transferred together with its complementary R gene to a given crop plant. To specifically trigger defense responses upon pathogen challenge, one of the two components is placed under the control of a pathogen-inducible plant promoter, whereas the other is constitutively expressed. Activation of the gene cassette will lead to localized HR that will prevent further spread of the invading pathogen. The effectiveness of this HR-mediated resistance, however, relies on the ability of the pathogen to induce the promoter, as well as on the timing of and sensitivity toward, the HR response. One of the major advantages of this strategy is to confer broad-spectrum disease resistance independent of AVR recognition and thus irrespective of the ability of pathogens to overcome R-gene-mediated resistance. This strategy has been used to create transgenic plants that show broad-spectrum and high-level fungal control (Stuiver and Custers, 2001). Since this strategy uses the endogenous defense components of the plant to engineer resistance, knowledge is required about how well downstream signaling components, including host-encoded virulence targets, are functionally conserved among plant species. *For example, R* genes frequently fail to function when transferred between plant species, especially when the species are not closely related. This suggests that the possibility to transfer resistance to commercially relevant crops by genetic engineering may be limited.

Transgenic tobacco plants expressing the elicitor protein cryptogein under the control of a pathogen-inducible promoter developed a HR in a normally compatible interaction with *P. parasitica* var. *nicotiana* (Keller et al., 1999). These transgenic plants also displayed enhanced resistance to fungal pathogens that were unrelated to *Phytophthora* species (Keller et al., 1999). Furthermore, the oomycete *infl* and fungal *Avr9* genes confer avirulence to PVX on tobacco and *Cf-9* tomato, respectively (Kamoun, Honée, et al., 1999). These results demonstrate that *Avr* genes can induce resistance to unrelated pathogens by the induction of a HR.

CONCLUSIONS AND FUTURE DIRECTIONS

To date, 11 *Avr* genes of fungal origin have been cloned and demonstrated to govern either host genotype or species specificity. For some of these *Avr* genes, sequence or structural homology was found to described gene (products), thereby suggesting a putative intrinsic function for the pathogen. In the case of AVR9 and AVR-Pita, which exhibit homology to a carboxypeptidase inhibitor and a metalloprotease, respectively, no biochemical evidence is yet available confirming these putative functions. On the other hand, AVR4 exhibits structural homology to invertebrate chitin-binding domain proteins, binds to chitin *in vitro,* and protects *T. viride,* and possibly *C. fulvum,* against cell wall disassembly by plant chitinases. In spite of the proposed contribution of several *Avr* genes to virulence of the fungus, a virulence role has only been assigned for the proteins encoded by *Ecp1, Ecp2,* and *NIP1.* It is conceivable that *Avr* genes exhibit a role in virulence that is either difficult to quantify or is functionally compensated for by other effector proteins. One interesting field for future research would therefore be to create near-isogenic strains by gene-replacement or gene-silencing that differ only at the *Avr* loci, allowing subsequent detailed comparisons on susceptible hosts to assess their contribution to virulence.

Recognition of AVR proteins is mediated by structurally different classes of R proteins. Detailed analysis of gene-for-gene pairs has provided further insight into how and where AVR proteins are recognized by resistant plant genotypes. Mutations in motifs that target AVR or R proteins to specific cellular compartments usually abolish AVR recognition by the corresponding *R* gene, suggesting that components required for triggering of defense responses are part of the same complex. This implies that recognition of extracytoplasmic proteins would occur at the extracellular side of plant plasmamembranes. Indeed, export of the CLV3 protein to the extracellular space of tomato is required for its interaction with and activation of the membrane-bound CLV1/CLV2 receptor complex. Yet in the case of the extracellular AVR and ECP proteins of *C. fulvum*, for which the location allows an interaction with the extracellular LRRs of the Cf proteins, thus far no such interaction has been demonstrated. Moreover, evidence for direct interaction between AVR and R proteins is very limited and has been demonstrated only between AvrPto and Pto, and between AVR-Pita and Pi-ta.

For most other gene-for-gene pairs, models have been proposed that describe an indirect rather than a direct perception of AVR proteins by R proteins. Consistent with these models is the involvement of, at least, a third component in the perception complex. Such a component might represent a putative virulence target (such as the AVR9-HABS) encoded by the host, which is guarded by the R protein. AVR proteins, however, might have more than one virulence target. This has been proposed for AvrPto, since mutants have been identified which retained their virulence function while their interaction with Pto was abolished, as well as for AVR4, for which binding to chitin is not required for induction of *Cf-4*-mediated resistance. Thus, in order to elucidate the intrinsic function of AVR proteins, it is important to identify their putative virulence targets. Indirect perception, on the other hand, might also involve coreceptors (such as protein kinases) that form a complex with R proteins. Ligand-induced conformational changes or phosphorylation of receptors could facilitate transmission from the extracytoplasmic to the cytoplasmic domains and subsequently trigger signal transduction. Moreover, indirect perception could also involve protease-dependent elicitation of defense responses, which has been proposed for AVR-Pita, which encodes a putative metalloprotease, as well as for AVR2, which requires a cyteine endoprotease Rcr3 to trigger *Cf-2*-mediated defense responses. Ongoing research should identify additional players involved in the current models of AVR perception by disease resistant plants.

Several *R* genes of the NB-LRR family have been cloned which confer resistance toward *M. lini* (*L6, M, N,* and *P*), *P. parasitica* (*RPP1, 4, 5, 8, 10, 13,* and *14*), *B. lactucae* (*Dm3*), *V. dahliae* (*Ve1* and *Ve2*), *F. oxysporum* (*I2*), and *P. infestans* (*R1*). Most of these fungal and oomycetous pathogens, however, are not very amenable to molecular manipulation, which hampers cloning of the matching *Avr* genes. Although the NB-LRR proteins lack any obvious subcellular targeting signatures, in at least one NB-LRR protein (RPM1), which recognizes intracellular targeted AvrRpm1 and AvrB, association with the plasmamembrane as a result of myristylation has been demonstrated (Boyes et al., 1998). The putative cytoplasmic localization of the NB-LRR proteins suggests intracellular perception of the *Avr* gene products. Most fungal pathogens that are recognized by these NB-LRR proteins, however, produce haustoria and grow extracellularly. This implies that these pathogens might use specialized mechanisms for delivery of AVR proteins into plant cells. Whether such mechanisms show parallels with the bacterial *Hrp*-mediated type III secretion system (Bonas and Van den Ackerveken, 1997) needs to be elucidated.

The maintenance of *Avr* genes in pathogen populations suggests that the primary function of these genes is to contribute to virulence of the pathogen. A focus for future research should involve breeding strategies aimed at plant resistance directed against essential effector proteins of pathogens, which will lead to development of crop species that harbor durable resistance. However, to achieve such durability, more insight is required into the molecular mechanisms underlying the adaptation of pathogens to *R*-gene-mediated resistance.

REFERENCES

Anderson, P.A., Okubara, P.A., Arreyogarcia, R., Meyers, B.C., and Michelmore, R.W. (1996). Molecular analysis of irradiation-induced and spontaneous deletion mutants at a disease resistance locus in *Lactuca sativa. Molecular General Genetics* 251: 316-325.

Ballvora, A., Ercolano, M.R., Weiss, J., Meksem, K., Bormann, C.A., Oberhagemann, P., Salamini, F., and Gebhardt, C. (2002). The *R1* gene for potato resistance to late blight *(Phytophthora infestans)* belongs to the leucine zipper/NBS/LRR class of plant resistance genes. *The Plant Journal* 30: 361-371.

Biffen, R.H. (1905). Mendel's law of inheritance and wheat breeding. *Journal of Agricultural Science* 1: 4-48.

Bittner-Eddy, P.D., Crute, I.R., Holub, E.B., and Beynon, J.L. (2000). *RPP13* is a simple locus in *Arabidopsis thaliana* for alleles that specify downy mildew resistance to different avirulence determinants in *Peronospora parasitica. The Plant Journal* 21: 177-188.

Boissy, G., De La Fortelle, E., Kahn, R., Huet, J.C., Bricogne, G., Pernollet, J.C., and Brunie, S. (1996). Crystal structure of a fungal elicitor secreted by *Phytophthora cryptogea,* a member of a novel class of plant necrotic proteins. *Structure* 4: 1429-1439.

Bonas, U. and Lahaye, T. (2002). Plant disease resistance triggered by pathogen-derived molecules: Refined models of specific recognition. *Current Opinion in Microbiology* 5: 44-50.

Bonas, U. and Van den Ackerveken, G.F. (1997). Recognition of bacterial avirulence proteins occurs inside the plant cell: A general phenomenon in resistance to bacterial diseases? *The Plant Journal* 12: 1-7.

Botella, M.A., Parker, J.E., Frost, L.N., Bittner-Eddy, P.D., Beynon, J.L., Daniels, M.J., Holub, E.B., and Jones, J.D.G. (1998). Three genes of the *Arabidopsis RPP1* complex resistance locus recognize distinct *Peronospora parasitica* avirulence determinants. *Plant Cell* 10: 1847-1860.

Bourque, S., Binet, M.N., Ponchet, M., Pugin, A., and Lebrun-Garcia, A. (1999). Characterization of the cryptogein binding sites on plant plasma membranes. *Journal of Biological Chemistry* 274: 34699-34705.

Boyes, D.C., Nam, J., and Dangl, J.L. (1998). The *Arabidopsis thaliana RPM1* disease resistance gene product is a peripheral plasma membrane protein that is degraded coincident with the hypersensitive response. *Proceedings of the National Academy of Sciences, USA* 95: 15849-15854.

Bryan, G.T., Wu, K., Farrall, L., Jia, Y., Hershey, H.P., McAdams, S.A., Faulk, K.N., Donaldson, G.K., Tarchini, R., and Valent, B. (2000). A single amino acid difference distinguishes resistant and susceptible alleles of the rice blast resistance gene *Pi-ta. Plant Cell* 12: 2033-2046.

Chang, J.H., Tobias, C.M., Staskawicz, B.J., and Michelmore, R.W. (2001). Functional studies of the bacterial avirulence protein AvrPto by mutational analysis. *Molecular Plant-Microbe Interactions* 14: 451-459.

Colas, V., Conrod, S., Venard, P., Keller, H., Ricci, P., and Panabières, F. (2001). Elicitin genes expressed in vitro by certain tobacco isolates of *Phytophthora parasitica* are down regulated during compatible interactions. *Molecular Plant-Microbe Interactions* 14: 326-335.

Dangl, J.L. and Jones, J.D.G. (2001). Plant pathogens and integrated defense responses to infection. *Nature* 411: 826-833.

Day, P.R. (1957). Mutation to virulence in *Cladosporium fulvum. Nature* 179: 1141-1142.

De Jong, C.F., Honée, G., Joosten, M.H.A.J., and De Wit, P.J.G.M. (2000). Early defense responses induced by AVR9 and mutant analogues in tobacco cell suspensions expressing the *Cf-9* resistance gene. *Physiological and Molecular Plant Pathology* 56: 169-177.

De Jong, C.F., Takken, F.L.W., Cai, X., De Wit, P.J.G.M., and Joosten, M.H.A.J. (2002). Attenuation of *Cf*-mediated defense responses at elevated temperatures correlates with a decrease in elicitor-binding sites. *Molecular Plant-Microbe Interactions* 15: 1040-1049.

De Wit, P.J.G.M. (1992). Molecular characterisation of gene-for-gene systems in plant-fungus interactions and the application of avirulence genes in control of plant pathogens. *Annual Review of Phytopathology* 30: 391-418.

Dixon, M.S., Golstein, C., Thomas, C.M., Van der Biezen, E.A., and Jones, J.D.G. (2000). Genetic complexity of pathogen perception by plants: The example of

Rcr3, a tomato gene required specifically by *Cf-2*. *Proceedings of the National Academy of Sciences, USA* 97: 8807-8814.

Dixon, M.S., Hatzixanthis, K., Jones, D.A., Harrison, K., and Jones, J.D.G. (1998). The tomato *Cf-5* disease resistance gene and six homologs show pronounced allelic variation in leucine rich repeat copy number. *Plant Cell* 10: 1915-1925.

Dixon, M.S., Jones, D.A., Keddie, J.S., Thomas, C.M., Harrison, K., and Jones, J.D.G. (1996). The tomato *Cf-2* disease resistance locus comprises two functional genes encoding leucine-rich repeat proteins. *Cell* 84: 451-459.

Farman, M.L., Eto, Y., Nakao, T., Tosa, Y., Nakayashiki, H., Mayama, S., and Leong, S.A. (2002). Analysis of the structure of the *AVR1-CO39* avirulence locus in virulent rice-infecting isolates of *Magnaporthe grisea*. *Molecular Plant-Microbe Interactions* 15: 6-16.

Farrer, W. (1898). The making and improvement of wheats for Australian conditions. *Agricultural Gazette NSW* 9: 131-168.

Flor, H.H. (1942). Inheritance of pathogenicity in *Melampsora lini*. *Phytopathology* 32: 653-669.

Flor, H.H. (1958). Mutation to wider virulence in *Melampsora lini*. *Phytopathology* 48: 297-301.

Halterman, D., Zhou, F., Wei, F., Wise, R.P., and Schulze-Lefert, P. (2001). The MLA6 coiled-coil, NBS-LRR protein confers AvrMla6-dependent resistance specificity to *Blumeria graminis* f. sp. *hordei* in barley and wheat. *The Plant Journal* 25: 335-348.

Heath, M.C. (2000). Nonhost resistance and nonspecific plant defenses. *Current Opinion in Plant Biology* 3: 315-319.

Huet, J.C., Le Caer, J.P., Nespoulous, C., and Pernollet, J.C. (1995). The relationship between the toxicity and the primary and secondary structures of elicitin-like protein elicitors secreted by the phytopathogenic fungus *Pythium vexans*. *Molecular Plant-Microbe Interactions* 8: 302-310.

Idnurm, A. and Howlett, B.J. (2001). Pathogenicity genes of phytopathogenic fungi. *Molecular Plant Pathology* 2: 241-255.

Islam, M.R. and Mayo, G.M.E. (1990). A compendium on host genes in flax conferring resistance to flax rust. *Plant Breeding* 104: 89-100.

Jia, Y., McAdams, S.A., Bryan, G.T., Hershey, H.P., and Valent, J.C. (2000). Direct interaction of resistance gene and avirulence gene products confers rice blast resistance. *EMBO Journal* 19: 4004-4014.

Johnson, T., Newton, M., and Brown, A.M. (1934). Further studies of the inheritance of spore colour and pathogenicity in crosses between physiologic forms of *Puccinia graminis* f. sp. *tritici*. *Scientific Agriculture* 14: 360-373.

Jones, D.A. and Jones, J.D.G. (1996). The roles of leucine-rich repeats in plant defenses. *Advances in Botanical Research/Advances in Plant Pathology* 24: 90-167.

Jones, D.A., Thomas, C.M., Hammond-Kosack, K.E., Balint-Kurti, P.J., and Jones, J.D.G. (1994). Isolation of the tomato *Cf-9* gene for resistance to *Cladosporium fulvum* by transposon tagging. *Science* 266: 789-793.

Joosten, M.H.A.J., Cozijnsen, T.J., and De Wit, P.J.G.M. (1994). Host resistance to a fungal tomato pathogen lost by a single base-pair change in an avirulence gene. *Nature* 367: 384-386.

Joosten, M.H.A.J. and De Wit, P.J.G. M. (1999). The tomato-*Cladosporium fulvum* interaction: A versatile experimental system to study plant-pathogen interactions. *Annual Review of Phytopathology* 37: 335-367.

Joosten, M.H.A.J., Verbakel, H.M., Nettekoven, M.E., Van Leeuwen, J., Van der Vossen, R.T.M, and De Wit, P.J.G.M. (1995). The phytopathogenic fungus *Cladosporium fulvum* is not sensitive to the chitinase and β-1,3-glucanase defense proteins of its host tomato. *Physiological and Molecular Plant Pathology* 46: 45-59.

Joosten, M.H.A.J., Vogelsang, R., Cozijnsen, T.J., Verberne, M.C., and De Wit, P.J.G.M. (1997). The biotrophic fungus *Cladosporium fulvum* circumvents *Cf-4*-mediated resistance by producing unstable AVR4 elicitors. *Plant Cell* 9: 367-379.

Kamoun, S., Honée, G., Weide, R., Laugé, R., Kooman-Gersmann, M., De Groot, K., Govers, F., and De Wit, P.J.G.M. (1999). The fungal gene *Avr9* and the oomycete gene *inf1* confer avirulence to potato virus X on tobacco. *Molecular Plant-Microbe Interactions* 12: 459-462.

Kamoun, S., Huitema, E., and Vleeshouwers, V.G. (1999). Resistance to oomycetes: A general role for the hypersensitive response? *Trends in Plant Science* 4: 196-200.

Kamoun, S., Van West, P., De Jong, A.J., De Groot, K.E., Vleeshouwers, V.G., and Govers, F. (1997). A gene encoding a protein elicitor of *Phytophthora infestans* is down-regulated during infection of potato. *Molecular Plant-Microbe Interactions* 10: 13-20.

Kamoun, S., Van West, P., Vleeshouwers, V.G., De Groot, K.E., and Govers, F. (1998). Resistance of *Nicotiana benthamiana* to *Phytophthora infestans* is mediated by the recognition of the elicitor protein INF1. *Plant Cell* 10: 1413-1426.

Kang, S., Lebrun, M.H., Farrall, L., and Valent, B. (2001). Gain of virulence caused by insertion of a Pot3 transposon in a *Magnaporthe grisea* avirulence gene. *Molecular Plant-Microbe Interactions* 14: 671-674.

Kang, S., Sweigard, J.A., and Valent, B. (1995). The *PWL* host specificity gene family in the blast fungus *Magnaporthe grisea*. *Molecular Plant-Microbe Interactions* 8: 939-948.

Kawchuk, L.M., Hachey, J., Lynch, D.R., Kulcsar, F., Van Rooijen, G., Waterer, D.R., Robertson, A., Kokko, E., Byers, R., Howard, R.J., Fischer, R., and Prufer, D. (2001). Tomato *Ve* disease resistance genes encode cell surface-like receptors. *Proceedings of the National Academy of Sciences, USA* 98: 6511-6515.

Keller, H., Pamboukdjian, N., Ponchet, M., Poupet, A., Delon, R., Verrier, J.L., Roby, D., and Ricci, P. (1999). Pathogen-induced elicitin production in transgenic tobacco generates a hypersensitive response and nonspecific disease resistance. *Plant Cell* 11: 223-235.

Kooman-Gersmann, M., Honée, G., Bonnema, G., and De Wit, P.J.G.M. (1996). A high-affinity binding site for the AVR9 peptide elicitor of *Cladosporium fulvum* is present on plasma membranes of tomato and other solanaceous plants. *Plant Cell* 8: 929-938.

Kooman-Gersmann, M., Vogelsang, R., Hoogendijk, E.C., and De Wit, P.J.G.M. (1997). Assignment of amino acid residues of the AVR9 peptide of *Cladosporium fulvum* that determine elicitor activity. *Molecular Plant-Microbe Interactions* 10: 821-829.

Krüger, J., Thomas, C.L., Golstein, C., Dixon, M. S., Smoker, M., Tang, S., Mulder, L., and Jones, J.D.G. (2002). A tomato cysteine protease required for *Cf-2*-dependent disease resistance and suppression of autonecrosis. *Science* 296: 744-747.

Laugé, R. and De Wit, P.J.G.M. (1998). Fungal avirulence gene: Structure and possible functions. *Fungal Genetics and Biology* 24: 285-297.

Laugé, R., Goodwin, P.H., De Wit, P.J.G.M., and Joosten, M.H.A.J. (2000). Specific HR-associated recognition of secreted proteins from *Cladosporium fulvum* occurs in both host and nonhost plants. *The Plant Journal* 23: 735-745.

Laugé, R., Joosten, M.H.A.J., Haanstra, J.P., Goodwin, P.H., Lindhout, P., and De Wit, P.J. G.M. (1998). Successful search for a resistance gene in tomato targeted against a virulence factor of a fungal pathogen. *Proceedings of the National Academy of Sciences, USA* 95: 9014-9018.

Laugé, R., Joosten, M.H.A.J., Van den Ackerveken, G.F.J.M., Van den Broek, H.W., and De Wit, P.J.G.M. (1997). The in planta-produced extracellular proteins ECP1 and ECP2 of *Cladosporium fulvum* are virulence factors. *Molecular Plant-Microbe Interactions* 10: 725-734.

Luck, J.E., Lawrence, G.J., Dodds, P.N., Shepherd, K.W., and Ellis, J.G. (2000). Regions outside of the leucine-rich repeats of flax rust resistance proteins play a role in specificity determination. *Plant Cell* 12: 1367-1377.

Luderer, R., De Kock, M.J.D., Dees, R.H.L., De Wit, P.J.G.M., and Joosten, M.H.A.J. (2002). Functional analysis of cysteine residues of ECP elicitor proteins of the fungal tomato pathogen *Cladosporium fulvum*. *Molecular Plant Pathology* 3: 91-95.

Luderer, R., Rivas, S., Nürnberger, T., Mattei, B., Van den Hooven, H.W., Van der Hoorn, R.A.L., Romeis, T., Wehrfritz, J.M., Blume, B., Nennstiel, D., et al. (2001). No evidence for binding between resistance gene product Cf-9 of tomato and avirulence gene product AVR9 of *Cladosporium fulvum*. *Molecular Plant-Microbe Interactions* 14: 867-876.

Luderer, R., Takken, F.L.W., De Wit, P.J.G.M., and Joosten, M.H.A.J. (2002). *Cladosporium fulvum* overcomes *Cf-2*-mediated resistance by producing truncated AVR2 elicitor proteins. *Molecular Microbiology* 45: 875-884.

Mackey, D., Holt, B.F. III, Wiig, A., and Dangl, J.L. (2002). RIN4 interacts with *Pseudomonas syringae* type III effector molecules and is required for RPM1-mediated resistance in *Arabidopsis*. *Cell* 108: 743-754.

Marmeisse, R., Van den Ackerveken, G.F.J.M., Goosen, T., De Wit, P.J.G.M., and Van den Broek, H.W.J. (1993). Disruption of the avirulence gene *Avr9* in the two races of the tomato pathogen *Cladosporium fulvum* causes virulence on the to-

134 *FUNGAL DISEASE RESISTANCE IN PLANTS*

mato genotypes with the complementary resistance gene *Cf-9*. *Molecular Plant-Microbe Interactions* 6: 412-417.

Martin, G.B., Brommonschenkel, S.H., Chunwongse, J., Frary, A., Ganal, M.W., Spivey, R., Wu, T., Earle, E.D., and Tanksley, S.D. (1993). Map-based cloning of a protein kinase gene conferring disease resistance in tomato. *Science* 262: 1432-1436.

McDowell, J.M., Cuzick, A., Can, C., Beynon, J., Dangl, J.L., and Holub, E.B. (2000). Downy mildew *(Peronospora parasitica)* resistance genes in *Arabidopsis* vary in functional requirements for NDR1, EDS1, NPR1 and salicylic acid accumulation. *The Plant Journal* 22: 523-529.

McDowell, J.M., Dhandaydham, M., Long, T.A., Aarts, M.G., Goff, S., Holub, E.B., and Dangl, J.L. (1998). Intragenic recombination and diversifying selection contribute to the evolution of downy mildew resistance at the *RPP8* locus of *Arabidopsis*. *Plant Cell* 10: 1861-1874.

Mikes, V., Milat, M.L., Ponchet, M., Panabières, F., Ricci, P., and Blein, J.-P. (1998). Elicitins secreted by *Phytophthora* are a new class of sterol carrier proteins. *Biochemistry and Biophysics Research Communication* 245: 133-139.

Nürnberger, T. (1999). Signal perception in plant pathogen defence. *Cellular and Molecular Life Science* 55: 167-182.

Oort, A.J.P. (1944) Onderzoekingen over stuifbrand: II. Overgevoeligheid voor stuifbrand *(Ustilago tritici)*. *Planteziekten* 50: 73-106.

Orbach, M.J., Farrall, L., Sweigard, J.A., Chumley, F.G., and Valent, B. (2000). A telomeric avirulence gene determines efficacy for the rice blast resistance gene *Pi-ta*. *Plant Cell* 12: 2019-2032.

Ori, N., Eshed, Y., Paran, I., Presting, G., Aviv, D.H., Tanksley, S.D., Zamir, D., and Fluhr, D. (1997). The *I2C* family from the wilt disease resistance locus *I2* belongs to the nucleotide binding, leucine-rich repeat superfamily of plant resistance genes. *Plant Cell* 9: 521-532.

Osman, H., Vauthrin, S., Mikes, V., Milat, M.L., Panabieres, F., Marais, A., Brunie, S., Maume, B., Ponchet, M., and Blein, J.P. (2001). Mediation of elicitin activity on tobacco is assumed by elicitin-sterol complexes. *Molecular Biology of the Cell* 12: 2825-2834.

Parniske, M., Hammond-Kosack, K.E., Golstein, C., Thomas, C.M., Jones, D.A., Harrison, K., Wulff, B.B., and Jones, J.D.G. (1997). Novel disease resistance specificities result from sequence exchange between tandemly repeated genes at the *Cf-4/9* locus of tomato. *Cell* 91: 821-832.

Pérez-García, A., Snoeijers, S.S., Joosten, M.H.A.J., Goosen, T., and De Wit, P.J.G.M. (2001). Expression of the avirulence gene *Avr9* of the fungal tomato pathogen *Cladosporium fulvum* is regulated by the global nitrogen response factor NRF1. *Molecular Plant-Microbe Interactions* 14: 316-325.

Ricci, P., Trentin, F., Bonnet, P., Venard, P., Mouton-Perronet, F., and Bruneteau, M. (1992). Differential production of parasiticein, an elicitor of necrosis and resistance in tobacco, by isolates of *Phytophthora parasitica*. *Plant Pathology* 41: 298-307.

Richter, T.E., Pryor, T. J., Bennetzen, J.L., and Hulbert, S.H. (1995). New rust resistance specificities associated with recombination in the *Rp1* complex in maize. *Genetics* 141: 373-381.

Ritter, C. and Dangl, J.L. (1996). Interference between two specific pathogen recognition events mediated by distinct plant disease resistance genes. *Plant Cell* 8: 251-257.

Rivas, S., Mucyn, T., Van den Burg, H.A., Vervoort, J., and Jones, J.D.G. (2002). An approximately 400 kDa membrane-associated complex that contains one molecule of the resistance protein Cf-4. *The Plant Journal* 29: 1-16.

Rivas, S., Romeis, T., and Jones, J.D.G. (2002). The Cf-9 disease resistance protein is present in a ~420-kilodalton heterodimeric membrane-associated complex at one molecule per complex. *Plant Cell* 14: 1-15.

Rohe, M., Gierlich, A., Hermann, H., Hahn, M., Schmidt, B., Rosahl, S., and Knogge, W. (1995). The race-specific elicitor, NIP1, from the barley pathogen, *Rhynchosporium secalis,* determines avirulence on host plants of the *Rrs1* resistance genotype. *EMBO Journal* 14: 4168-4177.

Rojo, E., Sharma, V.K., Kovaleva, V., Raikhel, N.V., and Fletcher, J.C. (2002). CLV3 is localized to the extracellular space, where it activates the *Arabidopsis* CLAVATA stem cell signalling pathway. *Plant Cell* 14: 969-977.

Rose, J.K.C., Ham, K.-S, Darvill, A.G., and Albersheim, P. (2002). Molecular cloning and characterisation of glucanase inhibitor proteins (GIPs): Co-evolution of a counter-defense mechanism by plant pathogens. *Plant Cell* 14: 1329-1345.

Salmeron, J.M., Oldroyd, G.E.D., Rommens, C.M.T., Scofield, S.R., Kim, H.S., Lavelle, D.T., Dahlbeck, D., and Staskawicz, B.J. (1996). Tomato *Prf* is a member of the leucine-rich repeat class of plant disease resistance genes and lies embedded within the *Pto* kinase gene cluster. *Cell* 86: 123-133.

Scofield, S.R., Tobias, C.M., Rathjen, J.P., Chang, J.H., Lavelle, D.T., Michelmore, R.W., and Staskawicz, B.J. (1996). Molecular basis of gene-for-gene specificity in bacterial speck disease of tomato. *Science* 274: 2063-2065.

Staskawicz, B.J., Mudgett, M.B., Dangl, J.L., and Galan, J.E. (2001). Common and contrasting themes of plant and animal diseases. *Science* 292: 2285-2289.

Stephenson, S.A., Hatfield, J., Rusu, A.G., Maclean, D.J., and Manners, J.M. (2000). *CgDN3:* An essential pathogenicity gene of *Colletotrichum gloeosporioides* necessary to avert a hypersensitive-like response in the host *Stylosanthes guianensis. Molecular Plant-Microbe Interactions* 13: 929-941.

Stuiver, M.H. and Custers, J.H.H.V. (2001). Engineering disease resistance in plants. *Nature* 411: 865-868.

Sweigard, J.A., Carroll, A.M., Kang, S., Farrall, L., Chumley, F.G., and Valent, B. (1995). Identification, cloning, and characterization of *PWL2,* a gene for host species specificity in the rice blast fungus. *Plant Cell* 7: 1221-1233.

Swiderski, M.R. and Innes, R.W. (2001). The *Arabidopsis PBS1* resistance gene encodes a member of a novel protein kinase subfamily. *The Plant Journal* 26: 101-112.

Takken, F.L. and Joosten, M.H.A.J. (2000). Plant resistance genes: Their structure, function and evolution. *European Journal of Plant Pathology* 106: 699-713.

Takken, F.L., Luderer, R., Gabriels, S.H., Westerink, N., Lu, R., De Wit, P.J.G.M., and Joosten, M.H.A.J. (2000). A functional cloning strategy, based on a binary PVX-expression vector, to isolate HR-inducing cDNAs of plant pathogens. *The Plant Journal* 24: 275-283.

Takken, F.L., Thomas, C.M., Joosten, M.H.A.J., Golstein, C., Westerink, N., Hille, J., Nijkamp, H.J., De Wit, P.J.G.M., and Jones, J.D.G. (1999). A second gene at the tomato *Cf-4* locus confers resistance to *Cladosporium fulvum* through recognition of a novel avirulence determinant. *The Plant Journal* 20: 279-288.

Tang, X., Frederick, R.D., Zhou, J., Halterman, D.A., Jia, Y., and Martin, G.B. (1996). Initiation of plant disease resistance by physical interaction of AvrPto and Pto kinase. *Science* 274: 2060-2063.

Thomas, C.M., Jones, D.A., Parniske, M., Harrison, K., Balint-Kurti, P.J., Hatzixanthis, K., and Jones, J.D.G. (1997). Characterization of the tomato *Cf-4* gene for resistance to *Cladosporium fulvum* identifies sequences that determine recognitional specificity in Cf-4 and Cf-9. *Plant Cell* 9: 2209-2224.

Van den Ackerveken, G.F.J.M., Dunn, R.M., Cozijnsen, T.J., Vossen, J.P.M.J., Van den Broek, H.W.J., and De Wit, P.J.G.M. (1994). Nitrogen limitation induces expression of the avirulence gene *avr9* in the tomato pathogen *Cladosporium fulvum*. *Molecular General Genetics* 243: 277-285.

Van den Ackerveken, G.F.J.M., Van Kan, J.A.L., and De Wit, P.J.G.M. (1992). Molecular analysis of the avirulence gene *Avr9* of the fungal tomato pathogen *Cladosporium fulvum* fully supports the gene-for-gene hypothesis. *The Plant Journal* 2: 359-366.

Van den Ackerveken, G.F.J.M., Van Kan, J.A.L., Joosten, M.H.A.J., Muisers, J.M., Verbakel, H.M., and De Wit, P.J.G.M. (1993). Characterization of two putative pathogenicity genes of the fungal tomato pathogen *Cladosporium fulvum*. *Molecular Plant-Microbe Interactions* 6: 210-215.

Van den Hooven, H.W., Van den Burg, H.A., Vossen, P., Boeren, S., De Wit, P.J.G.M., and Vervoort, J. (2001). Disulfide bond structure of the AVR9 elicitor of the fungal tomato pathogen *Cladosporium fulvum:* Evidence for a cysteine knot. *Biochemistry* 40: 3458-3466.

Van der Biezen, E.A., Freddie, C.T., Kahn, K., Parker, J.E., and Jones, J.D.G. (2002). *Arabidopsis RPP4* is a member of the *RPP5* multigene family of *TIR-NB-LRR* genes and confers downy mildew resistance through multiple signalling components. *The Plant Journal* 29: 439-451.

Van der Biezen, E.A. and Jones, J.D.G. (1998). Plant disease-resistance proteins and the gene-for-gene concept. *Trends Biochemical Science* 23: 454-456.

Van der Biezen, E.A., Sun, J., Coleman, M.J., Bibb, M.J., and Jones, J.D.G. (2000). *Arabidopsis RelA/SpoT homologues implicate (p)ppGpp in plant signalling. *Proceedings of the National Academy of Sciences, USA* 97: 3747-3752.

Van der Horrn, R.A.L. (2001). The Cf-4 and Cf-9 resistance proteins of tomato. PhD thesis. Wageningen University, the Netherlands.

Van der Hoorn, R.A.L., De Wit, P.J.G.M., and Joosten, M.H.A.J. (2002). Balancing selection favors guarding resistance proteins. *Trends in Plant Science* 7: 67-71.

Van Kan, J.A.L., Van den Ackerveken, G.F.J.M., and De Wit, P.J.G.M. (1991). Cloning and characterisation of complementary DNA of avirulence gene *avr9* of the fungal pathogen *Cladosporium fulvum,* causal agent of tomato leaf mould. *Molecular Plant-Microbe Interactions* 4: 52-59.

Wevelsiep, L., Rüpping, E., and Knogge, W. (1993). Stimulation of barley plasmalemma H$^+$-ATPase by phytotoxic peptides from the fungal pathogen *Rhynchosporium secalis. Plant Physiology* 101: 297-301.

White, F.F., Yang, B., and Johnson, L.B. (2000). Prospects for understanding avirulence gene function. *Current Opinion in Plant Biology* 3: 291-298.

Wubben, J.P., Joosten, M.H.A.J., and De Wit, P.J.G.M. (1994). Expression and localization of two in planta induced extracellular proteins of the fungal tomato pathogen *Cladosporium fulvum. Molecular Plant-Microbe Interactions* 7: 516-524.

Xiao, F., Tang, X., and Zhou, J.M. (2001). Expression of 35S: *Pto* globally activates defense-related genes in tomato plants. *Plant Physiology* 126: 1637-1645.

Xiao, S., Ellwood, S., Calis, O., Patrick, E., Li, T., Coleman, M., and Turner, J.G. (2001). Broad-spectrum mildew resistance in *Arabidopsis thaliana* mediated by *RPW8. Science* 291: 118-120.

Chapter 5

Pathogenesis-Related Proteins and Their Roles in Resistance to Fungal Pathogens

Jayaraman Jayaraj
Ajith Anand
Subbaratnam Muthukrishnan

INTRODUCTION

In the course of evolution and adaptation to pathogen pressure in the environment, higher plants have developed thick cell walls made up of cellulose, pectin, lignin, etc., which act as physical barriers to an invading pathogen. These barriers represent the primary defense of plants. The secondary defense mechanism involves the constitutive expression of plant secondary metabolites (for example, phenolics, alkaloids, and saponins) that constitute part of the chemical defenses. However, the most complex defense reaction involves turning on a cascade of genes involved in plant-pathogen interactions. The inducible defenses include the following: production of reactive oxygen species, phytoalexins, cell wall components (callose, hydroxylproline-rich proteins, etc.), and another group of proteins called pathogenesis-related (PR) proteins. Extensive study of the hypersensitive reaction of tobacco plants infected with tobacco mosaic virus (TMV) led to the discovery of PR proteins. These proteins were first identified as inducible proteins in leaf extracts reacting hypersensitively

We thank the Kansas Wheat Commission, the Kansas Sorghum Commission, and the U.S. Wheat and Barley Scab Initiative for financial assistance. This is contribution #03-69-B from the Kansas Agricultural Experiment Station.

to TMV. Based on their increasing order of electrophoretic mobility, four new protein components (I, II, III, and IV) were identified in the hypersensitive plants that were absent in the water-inoculated tobacco plants (Van Loon and Van Kammen, 1970).

Selective extraction methods and high-resolution polyacrylamide gel electrophoresis revealed that these classes of proteins were not peculiar not only to virus infection, but also to many bacterial and fungal attacks, particularly in plants exhibiting a hypersensitive reaction. These proteins are mostly of low molecular weight, preferentially extracted at low pH, resistant to proteolysis, and localized predominantly in the intercellular spaces of leaves (Van Loon, 1990). The term *pathogenesis-related proteins* was abbreviated as PR proteins to designate the "proteins coded for by the host plant but induced only in pathological or related situations" (Antoniw et al., 1980, p. 79). Here, the terms "pathological or related situations" refer to all types of pathogens and even parasitic attack by insects, nematodes, and other higher forms of animals. More recent reports have shown the induction of PR proteins as a result of colonization by non-pathogenic/beneficial fungi and bacteria (Blilou et al., 2000; Yedidia et al., 2000; Zehnder et al., 2001; Coventry and Dubery, 2001). Abiotic stresses and disorders were not included in this definition, although noninfectious symptoms such as toxin-induced chlorosis and necrosis often trigger induction of certain PR proteins. PR proteins are also induced upon environmental stresses, by chemical elicitors, and at different developmental stages of the plant. The major criterion for being classified as a PR protein is that the protein should be novel and induced upon infection and should impede further pathogen progression, but not in all pathological conditions (Van Loon, 1990). Their induction only in pathological situations suggests (but does not prove) a role for these proteins in plant defense.

Local infection or hypersensitive reactions in a plant can lead to acquisition of systemically enhanced resistance to subsequent infection by various types of pathogens, and this type of systemically acquired resistance (SAR) is attributed to induction of PR proteins in tissues that are remote from the site of infection. Some other related situations in which PR proteins are induced include application of chemicals that mimic the effect of pathogen and wound responses. Application of salicylic acid (SA), which is known to be an essential intermediate of signal-transduction pathways, led to accumulation of PR proteins in tobacco (Gaffney et al., 1993). Mechanical wounding or insect feeding also led to accumulation of PR proteins (Krishnaveni et al., 1999).

CLASSIFICATION OF PR PROTEINS

The PR proteins are classified into 14 groups (Van Loon and Van Strien, 1999), and an additional three groups of PR proteins have been proposed recently (Van Loon, 2001) (Table 5.1). Although the earlier classification was made for tobacco and tomato, the present nomenclature system also accommodates PR proteins from other plant species. The families are numbered and the different members of the same family are assigned letters according to the order in which they were described. For naming a PR protein, it is necessary to obtain information at both nucleic acid and protein levels and the same is true for a stress-related protein falling within the definition of PR proteins.

PR-1 Family

The PR-1 family is often the most abundant group of proteins and is induced to very high levels upon infection (reaching up to 1 to 2 percent of the total leaf protein). PR-1 proteins have been isolated from practically all plant species studied. They are typically about 160 amino acids in length, although there is some variation in their sizes from different plant species. The PR-1 family in tobacco consists of three acidic polypeptides, namely PR-1a, PR-1b, and PR-1c, which have >90 percent amino acid sequence identity with one another and are also serologically related to PR-1 proteins from other plant species. PR-1s fall into two groups, one being acidic and the other basic (sequence similarity between the two groups is about 65 percent). The acidic PR-1 genes have no introns and the encoded proteins remain soluble at acidic pH (pH 3.0). Acidic PR-1s are predominantly found in the extracellular spaces and xylem elements of TMV-infected plants, but they were also found in vacuoles of some specialized leaf cells (Dixon et al., 1991). The basic PR proteins are

FUNGAL DISEASE RESISTANCE IN PLANTS

TABLE 5.1. Families of PR proteins

Family	Representative protein	Biochemical properties	Molecular size range
PR-1	Tobacco PR-1a	Unknown	15-17 kDa
PR-2	Tobacco PR-2	β-glucanase	30-41 kDa
PR-3	Tobacco P, Q	Chitinase type I,II,IV,V,VI,VII	35-46 kDa
PR-4	Tobacco R	Chitinase type I,II	13-14 kDa
PR-5	Tobacco S	Thaumatin-like	16-26 kDa
PR-6	Tomato inhibitor I	Proteinase inhibitor	8-22 kDa
PR-7	Tomato P_{69}	Endoproteinase	69 kDa
PR-8	Cucumber chitinase	Chitinase type III	30-35 kDa
PR-9	Tobacco "lignin-forming peroxidase"	Peroxidase (POC)	50-70 kDa
PR-10	Parsley "PR1"	Ribonuclease-like	18-19 kDa
PR-11	Tobacco class V chitinase	Chitinase, type I	40 kDa
PR-12	Radish Rs-AFP3	Ion transport (defensins)	5 kDa
PR-13	Arabidopsis TH12.1	Thionins	5-7 kDa
PR-14	Barley LTP4	Lipid transfer proteins	9 kDa
Proposed groups:			
PR-15		Oxalate oxidases	22-25 kDa
PR-16		Oxalate oxidase-like protein	100 kDa (hexamer)

Source: Modified from Van Loon and Van Strien, 1999.

also found in both the extracellular spaces and the vacuoles of plants. Curiously, basic PR-1 proteins of tobacco do not cross-react with the antibodies raised against acidic PR proteins of tobacco, suggesting that the genes encoding these two groups of PR-1 proteins diverged a long time ago. Indeed, evidence suggests that these two groups are differentially regulated (Memelink et al., 1990). The biological function of PR-1 proteins is not yet clearly established.

The PR-1 proteins were initially believed to be involved in virus resistance or in virus localization in tobacco. An interspecific hybrid between *Nicotiana glutinosa* and *N. debneyi* constitutively producing PR-1 protein was shown to be highly resistant to TMV (Ahl and Gianinazzi, 1982). However, results obtained using transgenic tobacco plants expressing PR-1a or PR-1b did not support a role for these proteins in enhancing resistance to TMV (Linthorst et al., 1989)

β-1,3-Glucanases (PR-2 Family)

The PR-2 family comprises primarily β-1,3-glucanases, which are widely distributed among plant species. Several PR proteins with β-1,3-glucanase activity have been detected in tobacco, and they have been classified into four groups based on amino acid sequence similarities (Leubner-Metzger and Meins, 1999). Class I has several related basic isoforms which accumulate predominantly in vacuoles. In contrast, the class II (PR-2a, PR-2b, and PR-2c) and class III (PR-2d) members are acidic and are secreted into the extracellular space. Another distinct anther-specific glucanse, tag1, has been placed in a fourth group (Leubner-Metzger and Meins, 1999). The specific activities and substrate specificities vary considerably among the members. For instance, the class II enzyme appears to be 50 to 250 times more active in degrading laminarin than the class II PR-2a and PR-2b and the class III PR-2d enzymes. Enough experimental evidence is available to support a defensive role for β-1,3-glucanases against fungal pathogens, especially in a synergistic interaction with plant chitinases. These glucanohydrolases can exert their antifungal activity in at least two different ways: either directly by degrading the cell walls of the pathogen or indirectly by promoting the release of cell wall–degradation products that can act as elicitors to trigger a wide range of defense reactions (Leubner-Metzger and Meins, 1999). Several reports have confirmed the in vitro antifungal activity of β-1,3-glucanases (especially the vacuolar isoforms) against various fungi (Velazhahan et al., 2003). The oligosaccharides released from the cell walls of the pathogen as a result of digestion by β-1,3-glucanases acted as elicitors in soybean (*Glycine max* L.) plants. They induced the accumulation of a phytoalexin, glyceollin, which curtailed infection by *Phytophthora megasperma* f. sp. *glycinea* (Sharp et al., 1984).

Chitinases (PR-3, -4, -8, and -11)

The families PR-3, -4, -8, and -11 are comprised mainly of chitinases and related proteins. Chitinases hydrolyze the β-1,4 linkages between *N*-acetylglucosamine residues of chitin, a structural polysaccharide of the cell wall of many fungi, as well as the cuticle of insects and egg shells of nematodes. Chitinases were the first pathogen-induced proteins to be identified and have been studied more extensively than the other groups. Many chitinases have been purified from plants and their genes have been cloned (Neuhaus, 1999). The enzyme is linked with the thinning of the growing hyphal tips of fungi, followed by a balloonlike swelling that eventually leads to a bursting of hyphae. A combination of chitinase and β-1,3-glucanase was demonstrated to be more effective than either enzyme alone against many fungi (Mauch et al., 1988). Some of the chitinases, however, are not antifungal. For instance, the class IIa chitinases of tobacco are not antifungal, whereas class I members are strongly antifungal. In wheat, resistance to leaf rust in cultivars Karee and Lr35 was partially attributed to high constitutive levels of chitinase and induced β-1,3-glucanase activities, which resulted from a hypersensitive defense (Anguelova-Merhar et al., 2001). The lack of protection observed in other cases might be attributed to the site of accumulation and enzyme targeting. Apart from the active role of chitinases in fungal cell wall lysis, the degradation products of the fungal cell wall, especially the oligomers, could serve as elicitors (Muthukrishnan et al., 2001). Several chitinase genes have been cloned from legumes. These genes are expressed in root hairs or nodules, and it has been proposed that they serve as receptors for chitooligosaccharides or as modulators of nodulation in legumes (Staehelin et al., 1994). A cloned Nod-factor receptor-like protein has a PR-11 chitinase-like domain and a cytosolic kinase domain (Kim et al., 1997), providing further evidence for the role of Nod factors as substrates for chitinases.

Thaumatin-Like Proteins (PR-5 Family)

The PR-5 or thaumatin-like proteins (TLPs) have a high degree of sequence similarity with each other and show immunological relationship with the sweet-tasting protein thaumatin, found in fruits of the West African shrub *Thaumatococcus daniellii* (Cornelissen et al., 1986). TLPs are not commonly detected in leaves of young healthy plants but rapidly accumulate to high levels in response to biotic or abiotic stresses. They are generally highly soluble proteins that are stable even at very low pH and are resistant to proteolysis. The extracellular TLPs are always acidic, whereas the vacuolar ones tend to be basic (Velazhahan et al., 1999). As with other PR proteins, there are two classes of TLPs. The larger class includes proteins with a size of about 22 to 26 kDa, and members of the smaller class are about 16 to 17 kDa. For example, rice has two classes of inducible TLPs; one is a 16 kDa TLP induced by *Pseudomonas syringae* (Reimmann and Dudler, 1993). Infection of rice plants with the sheath blight pathogen *Rhizoctonia solani* resulted in the induction of two TLPs of 24 and 25 kDa within one to two days after infection (Velazhahan et al., 1998). Reiss and Horstmann (2001) detected eight TLPs (TLP1-TLP8) by two-dimensional electrophoresis and N-terminal sequencing in barley leaves infected by *Drechslera teres*. Four of these were acidic and four were basic. A 17 kDa TLP was recently purified from sorghum leaves infected by *Fusarium moniliforme,* which strongly inhibited the mycelial growth of the same pathogen and *Trichoderma viride* at 10 µg level (Velazhahan et al., 2002). Although many TLPs are antifungal, their antifungal activity varies with the specific fungal genus (Vigers et al., 1991; Abad et al., 1996). At higher concentrations, TLPs can actively lyse fungal membranes, while at lower concentrations, they affect membrane permeability which can cause leakage of cell constituents and increase the uptake of other antifungal compounds. The TLPs are induced in response to viral, fungal, and bacterial infections. TLPs are also induced by nonpathogenic stress, such as osmotic stress, and by hormones and signal molecules such as abscisic acid (ABA), ethylene, SA, and jasmonic acid (JA) and by wounding (summarized in Velazhahan et al., 1999).

Proteinase Inhibitors (PR-6 Family)

Proteinase inhibitors (PIs) are stable defense proteins found in seeds whose expression is developmentally regulated; they are also induced in leaves upon attack by pests or pathogens. Although more than ten unrelated subclasses of proteinase inhibitors are known in plants, only a few of them have been shown to have defensive roles in plants. The 6 family of PR proteins has been created to accommodate the PIs with known defensive roles in plants. The first such inhibitors were the two PIs induced in tomato upon insect feeding (Hass et al., 1982). Induction of PIs in plants in response to microbial attack has also been observed in many plant systems. Increases in trypsin and chymotrypsin inhibitory activities in tomato leaves infected by *Phytophthora infestans* were detected, and this induction was stronger in resistant lines when compared to the susceptible ones (Peng and Black, 1976). Pathogens and pests invading plant tissues rely on a set of proteases as part of their virulence factors. These proteinases belong to four classes, namely serine-, cysteine-, aspartic-, and metalloproteinases. In parallel, plants have evolved genes encoding inhibitors that inactivate some of these proteinases and thus may reduce the ability of the pathogen or pest to digest host proteins, and therefore limit the availability of nitrogen source for the invader. The induction of subclasses of PIs in response to microbial infection has been reported in numerous plants (summarized in Heitz et al., 1999). The plant defensive capabilities of PIs against insect predation have been clearly demonstrated both by feeding experiments with synthetic diets (Ryan, 1990) and bioassays on transgenic plants (Hilder et al., 1987). The search for PIs that inactivate proteinases of pathogenic microbes has generated a large amount of data. A clear understanding of the roles of PIs would be possible through genetic engineering of both pathogen and host, either by knocking out microbial genes encoding secreted proteases or by overexpressing new isoforms of PIs in plants.

Peroxidases (PR-9 Family)

Peroxidases are key enzymes in the cell wall-building process, and it has been suggested that extracellular or wall-bound peroxidases would enhance plant resistance by the construction of a cell wall barrier that may impede pathogen ingression and spread (Ride, 1983). Even though many peroxidases are found in most plant species and are expressed constitutively, some isozymes appear to be inducible upon pathogen infection. An increase in peroxidase activity has been correlated with resistance of plants to various pathogens. In contrast, the peroxidase activity in susceptible interactions was delayed or not induced within the critical time period of early host response, allowing the pathogen to proliferate in host tissues (Chittoor et al., 1999). Apart from their role in cell wall strengthening through deposition of characteristic materials required for processes such as lignification, suberization, and polysaccharide cross-linking, peroxidases also are involved in a secondary reaction. The formation of reactive oxygen species occurs during deposition of these compounds into the cell walls by peroxidase activity, which is likely to be toxic to the pathogen. Alternatively, these active oxygen species may act as messengers which activate the plant defense responses that would contribute to resistance (Chittoor et al., 1999). However, experiments to determine the role of peroxidases in defense responses through transgenic studies, such as overexpression or antisense suppression, have remained inconclusive (see Chapter 7).

Oxalate Oxidases, Lipid Transfer Proteins, Thionins, and Plant Defensins (PR-12, -13, -14, -15, and -16)

The most recent classification of PR-protein groups also includes several plant defense peptides, including defensins (PR-12), thionins (PR-13), lipid transfer proteins (LTPs, PR-14), and plant oxalate oxidases and related proteins (PR-15 and -16), all of which have antimicrobial activity. The oxalate oxidases *(OxO)* from wheat and barley, which have close resemblance to the germinlike proteins, have been extensively studied. These genes were first detected and isolated from the cell walls of germinating seeds and were later demonstrated to be induced by fungal invasion (Lane 1994; Hurkman and

Tanaka, 1996). Oxalate oxidases catalyze the degradation of oxalic acid into carbon dioxide and hydrogen peroxide. They are oligomeric, water-soluble, protease-resistant, heat-stable and SDS-tolerant glycoproteins (Lane, 1994). Even though the exact mechanism of antimicrobial activity of these enzymes is not clearly understood to date, they are implicated in the generation of reactive oxygen species (ROS).

Some of the cysteine-rich antimicrobial peptides reported in plants have sizes ranging from 2 to 9 kDa, and some of them have been identified and classified as thionins, defensins, and lipid transfer proteins. Thionins are classified in a family of homologous peptides that include the purothionins and hordothionins isolated from wheat and barley, respectively, and their homologues occur from various taxa, such as viscotoxins, phoratoxins, and crambins (Garcia-Olmedo et al., 1989). The mature thionin peptides are generally 45 to 47 amino acids in length, and based on their amino acid sequence similarity, they can be classified into at least five types (I-V). The exact biological role of thionins as defense genes is not well understood (Bohlmann, 1999), but there are reports of transgenic plants overexpressing thionins and viscotoxins with enhanced resistance to fungal pathogens (Carmona et al., 1993; Epple et al., 1997).

Plant defensins are also a family of cysteine-rich antimicrobial peptides, 45 to 54 amino acid residues in length, that are ubiquitous in the plant kingdom (reviewed by Conceição and Broekaert, 1999). They are structurally unrelated to thionins and have a cysteine-stabilized α, β motif. All known members of this family have eight disulfide-linked cysteines, including one at the C-terminus and a glycine residue. Plant defensins are preferentially located in peripheral cell layers and have also been reported to occur in the xylem, in stomatal cells, and in the cells that line the substomatal cavity, all of which are locations for pathogen entry (Broekaert et al., 1997). Rye seed defensins, Rs-AFP1 and Rs-AFP2, are prototype defensins with antifungal activity. Some defensins affect morphogenesis (cause branching of hyphae), whereas others merely inhibit hyphal growth. They seem to act by inducing rapid Ca^{++} uptake and K^+ efflux across fungal membranes (Thevissen et al., 1997). Some, but not all, defensins are inducible upon pathogen infection.

Plant lipid transfer proteins are 90 to 95 amino acid–long polypeptides that are secreted and accumulate in the cell walls and other exposed surfaces of infected plants. They exhibit antimicrobial activity in vitro (Garcia-Olmedo et al., 1995). The LTPs accumulate in outer cell layers of exposed surfaces (e.g., cell wall) at much higher concentrations than those required to inhibit the growth of pathogens in vitro (Kader, 1996; Velazhahan et al., 2001); because of their high isoelectric point, they may act as membrane permeabilizing agents. These proteins are generally secreted and externally associated with the cell, although a cytosolic role of these enzymes has been suggested (Sterk et al., 1991; Pyee et al., 1994). These proteins were so named because of their ability to stimulate the transfer of a broad range of lipids between membranes in vitro (Yamada, 1992). It has been proposed that these enzymes may be involved in secretion and deposition of extracellular lipophilic materials and in transport of cutin monomers required for biosynthesis of surface wax (Sterk et al., 1991). The role of LTPs as antimicrobial agents was demonstrated in vitro against *T. viride* and *R. solani* (Velazahan et al., 2002) and by in vivo expression of these proteins in transgenic tobacco and *Arabidopsis* (Molina and Garcia-Olmedo, 1997).

NONDEFENSE FUNCTIONS OF PR PROTEINS AND PR-LIKE PROTEINS

The classic definition of PR proteins as proteins expressed in "pathological or related situations" excludes several other proteins related in sequence to PR proteins if these proteins are expressed even at low levels in uninfected plants. The proteins expressed constitutively in healthy plants and others expressed during specific developmental stages, such as flowering, have been referred to as PR-like proteins (Van Loon, 1990). The PR and PR-like proteins may also have other important functions in addition to their role in plant defense, because of their tissue-specific expression, localization in apoplast and vacuoles, and because their synthesis is often influenced by endogenous and exogenous signals. Surprisingly, some PR proteins show similarity to many other non-PR proteins found in a variety of organisms. For instance, the PR-1 proteins show significant sequence similarity to proteins of diverse origin, such as insect venom of hornet wasp, mammalian sperm-coating glycoprotein, and certain proteins secreted by the fungus *Schizophyllum commune* during fruiting body production (Van Loon and Van Strien, 1999). Because PR proteins are often stress-inducible proteins, they could play a vital role in alleviating the harmful effects of these stress conditions and the consequent damage to the affected tissues. Invariably, the acidic and basic PR proteins were induced by agents/conditions such as ethylene, plasmolysis, drought, high salt, heat shock, UV, ABA, and JA, and by mechanical wounding or tissue damage as a result of insect feeding. There are also reports of induction of PR proteins in the absence of pathogenic attack, suggesting their involvement in protection of cellular structures, either physically by stabilizing sensitive membranes or macromolecules or chemically by keeping potentially harmful saprophytic microorganisms on tissue surfaces or in intercellular spaces under check (Van Loon and Van Strien, 1999).

The specificity of *Rhizobium*-plant interactions might be partly due to the active role played by the root chitinases, which utilize Nod factors as substrates (Staehelin et al., 1994). A specific chitinase with homology to PR-4 was essential for somatic embryogenesis of carrot cell suspensions (Kragh et al., 1996). The occurrence of PR proteins in floral parts suggests their specific physiological functions during flower development rather than in plant defense, which may be of secondary importance (Van Loon and Van Strien, 1999). These findings suggest that such developmentally regulated chitinases and glucanases could be implicated in the generation or turnover of signal molecules or morphogenic factors. Although there is a possibility that β-1,3-glucanases function as signal-generating enzymes acting on cell wall glucans, the natural substrates for chitinases in higher plant cells have not been identified. In addition, the developmentally regulated β-1,3-glucanases may have a role in microsporogenesis, such as dissolution of the callose wall of the tapetum-specific callose (β-1,3-glucans). Another possible role for β-1,3- and β-1,3-/β-1,4-glucanases is in cell wall degradation during seed germination (Leubner-Metzger and Meins, 1999).

Dual Cellular Location of PR Proteins

DNA sequence analysis and identification of open reading frames using sequence prediction software have indicated that many, if not all, of the PR proteins have signal peptide sequences (not signal anchors) at their N-termini, suggesting that these proteins are made on ribosomes attached to the endoplasmic reticulum (ER). It is very likely that the PR proteins are deposited in the lumen of the ER, where they are then transported to other locations, including secretory vesicles. Indeed, almost all of the acidic PR proteins have been identified in the apoplastic fluid of plant cells. This is true for PR-1, PR-2, PR-3, and PR-5 and possibly a few other classes as well. They are often referred to by the notation PR1a, PR2a, etc. Interestingly, in many cells, basic counterparts of all of these classes of acidic PR proteins are found to accumulate in the vacuoles. Sequence analyses of cDNA clones of the encoded basic PR proteins have indicated the presence of additional sequences at the C-termini. These C-terminal extensions have been shown to be necessary and sufficient for targeting of these proteins to the vacuoles (Neuhas et al., 1991; Neuhaus, 1999). In some cases, the vacuolar targeting signal may be found at the mature N-terminus. For example, deletion of a 16 amino acid–long N-terminus propeptide from the precursor of sweet potato PR protein, sporamin, caused the protein to be secreted instead of being targeted to the vacuoles (Matsuoka et al., 1990).

Many basic forms of PR proteins (intracellular) have been shown to have strong antifungal activity, in contrast to their acidic forms (extracellular), which are less toxic or even nontoxic to fungi. The fungal pathogens, biotrophs in particular, initially develop in the intercellular space and subsequently grow extracellularly in the necrotrophic phase. Thus, PR proteins that are accumulated predominantly in vacuoles (intracellular) practically have little effect on the fungal hyphae that rapidly proliferate in the intercellular spaces without penetrating the cell. Disruption of vacuoles will lead to liberation of hydrolytic enzymes to the intercellular spaces; this occurs at the late necrotic phase, when enough tissue damage would have already been caused (Vidhyasekaran, 1997).

NATURAL AND SYNTHETIC ELICITORS OF PR-PROTEIN GENES

Several molecules that are derived from pathogens or from host-pathogen interactions can serve as elicitors of PR-protein induction.

The elicitors derived from pathogens include oligosaccharides derived from chitin, glucan, and pectin, fungal cell wall derivatives, extracellular glycoproteins, polysaccharides, oligosaccharides, and harpins produced by pathogenic and nonpathogenic bacteria and fungi. Mechanical wounding, insect feeding, chemical elicitors, and environmental stresses also can switch on PR protein accumulation. Thus, the expression of PR-protein genes in plants is often influenced by multiple stimuli, but in contrast, no receptors have been unequivocally established for any of these signal molecules (Zhou, 1999).

Plants can respond to oligosaccharide elicitors produced during the process of infection. The involvement of oligosaccharides derived from β-glucan, chitin, chitosan, and pectin in signaling of plant defense machinery has been well established. These elicitors can cause induction of phytoalexins or PR-protein synthesis (Vidhyasekaran, 1997; Kuc, 2001). The lipopolysaccharides (LPS) from the cell walls of Gram-negative bacteria can influence the outcome of certain plant-pathogen interactions. The LPS of *Xanthomonas campestris* induced β-1,3-glucanase expression in *Brassica* and evoked a local induced response, whereas the LPS from *Escherichia coli* did not register any increase in β-1,3-glucanase levels (Newman et al., 2001). An LPS fraction purified from the biocontrol bacterium *Burkholderia cepacia,* was able to induce resistance against infection by *Phytophthora nicotianae* in tobacco. This LPS elicited the accumulation of a variety of PR proteins (at least six classes) in leaves (Coventry and Dubery, 2001). This could be one of the major mechanisms by which plant-growth-promoting rhizobacteria evoke induction of systemic resistance in plants, which is often potentiated by SA or JA (Newman et al., 2001; Ramamoorthy et al., 2001).

Another class of elicitors is typified by the polypeptides encoded by avirulence *(avr)* genes of the pathogen. A pathogen containing a particular *avr* gene is recognized by the host plant that carries a corresponding resistance *(R)* gene and activates disease resistance in the host. Over the past decade, extensive work has been done in this area and several *R* genes and their corresponding *avr* genes have been characterized (see Chapters 3 and 4). The recognition of the *avr* gene product by an *R* gene product normally activates a variety of defense responses that include PR-protein induction in plants (Zhou, 1999). A typical example of this phenomenon is an apoplast-located race-specific oligopeptide elicitor, encoded by the *avr9* gene from the pathogen *Cladosporium fulvum,* which causes leaf mold of tomato. When tomato plants were injected with the intercellular fluid containing this peptide, the transcription of β-1,3-glucanase and chitinase genes was activated in plants that carried the corresponding *R* gene but not in plants lacking this gene (Wooben et al., 1996).

Salicylic acid is an endogenous signal molecule that very often can induce PR proteins in plants. SA is postulated to bind to a receptor which, in turn, may trigger the signal transduction cascade leading to the production of transcription factors regulating PR-protein expression or other defense proteins (Metraux, 2001). In tobacco plants, SA induces the expression of at least nine different PR-protein genes, and interestingly, all of them were also induced by TMV infection. The endogenous levels of SA increased 50-fold in TMV-inoculated leaves followed by accumulation of PR proteins in resistant plants. Neither response was detected in susceptible plants. However, SA treatment of susceptible plants did lead to the triggering of PR-protein gene expression, suggesting that TMV infection could not enhance SA levels in susceptible plants and, therefore, failed to trigger PR-gene expression (Malamy et al., 1990). Tobacco plants expressing the *NahG* gene (salicylate hydroxylase gene from *Pseudomonas putida*) do not accumulate SA. These plants did not accumulate PR proteins upon TMV infection (Gaffney et al., 1993). Similarly, *Arabidopsis* plants expressing the *NahG* gene did not accumulate mRNAs encoding PR proteins, which emphasizes the role of SA in PR-protein induction (Lawton et al., 1994).

Jasmonic acid and methyl JA are the other signal compounds synthesized from linolenic acid that are commonly involved in stress responses of plants. JA induced the accumulation of several polypeptides, proteinase inhibitors, and ribosome-inactivating proteins (RIP) in many plants (reviewed in Vidhyasekaran, 1997). Methyl JA also induced the accumulation of a proteinase inhibitor in alfalfa that was wound inducible. A higher level of proteinase inhibitors was found in alfalfa plants following treatment with methyl JA (Farmer et al., 1992).

Exogenous application of chemicals, such as β-amino *n*-butanoic acid (BABA) or acibenzolar *S*-methyl (ASM), can induce the accumulation of PR proteins, especially when challenged with pathogens

(Siegrist et al., 2000; Ziadi et al., 2001; Silue et al., 2002). It has been proposed that these chemicals are involved in other signal trans-duction pathways. The *Arabidopsis nahG* mutant plants showed in-duced resistance following application of BABA. A similar enhance-ment of resistance was also observed with mutants with impaired ethylene or JA pathways (Zimmerli et al., 2000). The β-1,3-endo-glucanase mRNA levels were increased following treatment of citrus fruits with BABA (Porat et al., 2002). Similarly, BABA-treated cauli-flower seedlings challenged with *Peronospora parasitica* had a mas-sive increase of PR-2 proteins. Treatment with ASM resulted in a progressive increase in β-1,3-glucanase activity in these seedlings over several hours following inoculation with *P. parasitica* (Ziadi et al., 2001). Treatment of tobacco plants with BABA led to the induction of PR-1 mRNA that was concentration dependent, with the highest accumulation occurring at 40 μmol/L of BABA (Hae-Keun et al., 1999). But in the case of *Arabidopsis,* spraying of BABA even at a low concentration induced accumulation of PR proteins (Jacob et al., 2001). Various chemicals, including 2,6-dichloroisonicotinic acid (INA) and benzothiadiazole (BTH), have induced disease resistance in various plants, which was mostly due to systemic acquired resis-tance (SAR) and was associated with production of PR proteins (Oostendorp et al., 2001). Application of ASM to soil induced the ac-cumulation of PR proteins, including chitinase, β-1,3-glucanase, and TLP in sugarcane. These plants were fairly resistant to red rot infec-tion, caused by *Colletotrichum falcatum* (Rameshundar et al., 2000).

Ethylene is another endogenous regulator of PR-protein gene ex-pression and is often referred to as a stress-related phytohormone. Treatment of plants with ethylene caused an enhanced accumulation of PR proteins, including β-1,3 glucanases and chitinases (Ishige et al., 1991). Treatment of tobacco calli with 1 mM cobalt chloride, a well-known inhibitor of ethylene biosynthesis, resulted in inhibition of β-1,3-glucanase activity, and this inhibition was overcome by ex-posure to ethylene (Felix and Meins, 1987). Exposure of ozone-sensi-tive tobacco plants to ozone resulted in a significant increase in ethyl-ene production that was undetectable in tolerant plants. Prolonged ozone treatment for two days markedly increased the mRNA levels for β-1,3-glucanase and chitinase in sensitive plants, which provided evi-dence for the involvement of ethylene in the transcriptional induction of PR proteins (Ernst et al., 1992). Ethylene-mediated induction of PR proteins in tobacco follows two pathways, one of which is light dependent and the other which is not light dependent. The light dependence was demonstrated by growing plants in light/dark following ethylene treatment. Accumulation of PR-1b transcript levels in light was fivefold greater than in the dark (Eyal et al., 1992). In *Arabidopsis,* PR-1, PR-2, and PR-5 proteins were induced by ethylene (Lawton et al., 1994). However, Mauch and colleagues (1984) observed no down-regulation of synthesis of chitinase and β-1,3-glucanase in pea after treatment with aminoethoxyvinylglycine, a specific inhibitor of ethylene synthesis. Similarly, a mutant of tomato that constitutively overexpresses ethylene did not show any marked increases in the level of PR proteins (Belles et al., 1992). The reasons for these apparently contradictory results are unknown.

Systemin, a wound-inducible polypeptide detected in tomato leaves, triggered the synthesis of proteinase inhibitors I and II (Pearce et al., 1991). Homologues of systemin were found in potato and alfalfa. Prosystemin mRNA accumulated in the upper leaves of plants in which the lower leaves were wounded, demonstrating that prosystemin mRNA is systemically wound inducible. Tomato plants expressing antisense prosystemin transcripts exhibited complete suppression of the systemic wound induction of protease inhibitors (McGurl et al., 1992). This observation supports the vital role of systemins in signal transduction regulating the synthesis of protease inhibitors.

Sometimes the signal inducing PR-protein expression might be a very simple compound, such as a short peptide. The intercellular fluids of tobacco hybrids or a hypersensitive type that was infected by tobacco necrosis virus contained a signal molecule that was capable of inducing the synthesis of PR-1a and PR-1b in leaves of tobacco var. *Xanthi.* The intercellular fluid from uninfected plants did not elicit this response. The components of the intercellular fluid were resolved by molecular sieve chromatography and deduced to contain a small peptide, or perhaps an amino acid, that was not inactivated by proteases or boiling (Gordon-Weeks et al., 1991).

Apart from the effects of the signal molecules acting individually, there exists the possibility of synergism or antagonism among signal molecules on expression of PR-protein genes. For example, accumulation of osmotin (PR-5) mRNA was dramatically higher in tobacco leaf tissues when ethylene and SA were applied in combination compared to plants treated with either one alone. Similarly, a combination of SA and JA induced accumulation of PR-1b in 'Wisconsin 38' tobacco to a level that was severalfold greater compared to plants treated with SA alone (Xu et al., 1994).

MECHANISMS THAT PROTECT PATHOGENS FROM PR PROTEINS

Pathogens may produce enzymes or inhibitors that can protect them from lysis by cell wall–hydrolyzing enzymes of the host. Chitin deacetylase is produced by *Colletotrichum lindemuthianum* and- *C. lagenarium.* When cucumber plants were inoculated with *C. lagenarium,* chitin deacetylase activity increased severalfold, especially during lesion development. Such rapid deacetylation of chitin would render the polymer chain not susceptible to chitinase and remain intact in the form of partially *N*-acetylated chitosan polymers which remain bound to the hyphae and thus sustain the rigidity of the cell wall. This mechanism would partially protect the fungal cell wall from host chitinases (Seigrist and Kauss, 1990).

The pathogens may tend to play it safe by excluding chitin as a component of their cell wall, especially within the infection structures. The absence of chitin at the site of initial contact with the host plasma membrane is critical for the establishment of infection by intracellular biotrophs. Such infection structures devoid of chitin would be resistant to host chitinases. This would indirectly affect the induction of chitinases in host tissues because the degradation products of chitin are strong elicitors of chitinases. The lack of chitin in the primary infection structures was demonstrated by the use of gold-labeled wheat germ agglutinin (which binds to *N*-acetyl glucosamines in chitin) in combination with electron microscopy. The appressorial cones, infection peg, and young and intracellular hyphae of *C. lindemuthianum* remained unlabeled throughout the stages of infection, indicating the absence of chitin in these structures, whereas other structures, including conidia, appressoria, germinating conidia, and matured hyphae were labeled (O'Connell and Ride, 1990).

The slower accumulation of PR proteins at the site of infection during the early stages of infection would favor pathogenesis substantially. This is apparently the situation in susceptible plant types. *Brassica nigra,* which is resistant to *Phoma lingam,* showed a very rapid induction of PR-2 proteins and two acidic chitinases, whereas in the susceptible species *B. napus,* the induction of these PR proteins was delayed or absent. It has been suggested that there was a correlation between accumulation of PR proteins and the outcome of the infection (Dixelius, 1994).

In many fungal infections, the pathogen releases small amounts of elicitors from its cell wall that may be below the threshold for activating the defense signaling of the host. An incompatible strain of *Fusarium oxysporum* f. sp. *apii* liberated more chitin oligosaccharides than a compatible strain infecting celery roots. This reduction in the level of elicitors released by the compatible strain was associated with lower levels of PR-protein accumulation (compared to the incompatible strain), which was presumably favorable for infection (Krebs and Grumet, 1993).

The PR proteins are also known to undergo degradation in plant tissues that are undergoing senescence. Often this is the signal for opportunistic fungi or pathogens to initiate infection in these tissues. An aspartyl proteinase that is constitutively present in tomato could cleave the PR proteins. The fungal infection offers a conducive environment for the protease to be active in the apoplast by the lowering of pII (Rodrigo et al., 1989). Even in the absence of infection, mere elicitor treatment of plant tissues would lead to rapid degradation of PR proteins by native proteases. SA-treated tobacco leaf discs gradually get depleted of PR proteins with increases in time (Matsuoka and Ohashi, 1986).

A gradual adaptation of the pathogen to PR proteins in the host is the other type of escape mechanism often suspected, especially when overexpression of the native PR-protein genes in the same host is attempted. It is a possible mechanism by which transgenic plants constitutively overexpressing a PR-protein gene may not exhibit adequate levels of resistance to infection and disease. Transgenic tobacco plants overexpressing a basic chitinase in large quantities did not show resistance to frog eye spot pathogen. It might be due to the adaptation of the fungal pathogen to the host defense mechanism (Neuhaus et al., 1991).

TRANSGENIC PLANTS EXPRESSING SINGLE GENES FOR PR PROTEINS

It becomes a matter of curiosity for many researchers to determine the function of these PR genes when overexpressed in plants and their effect in controlling disease. Apart from that, it is an ideal strategy to enhance the level of resistance to pests. Under the aegis of advancements made in genetic engineering, tools are readily available to mobilize genes into plant systems and determine how they behave and function. It is also possible to target the expression and thereby ensure the PR proteins would be deployed in the right location where the pathogen initiates infection, e.g., intercellular space (Punja, 2001).

PR-1

Transgenic *N. tabacum* cv. Xanthi-nc constitutively overexpressing PR-1a showed a significant delay and substantial reduction in disease symptoms against two oomycetes, *Phytophthora parasitica* and *Peronospora tabacina,* suggesting their selective specificity to pathogens. Consistent with these results was the finding that purified PR-1 proteins inhibited the growth of *Phytophthora infestans* in in vitro antifungal assays (Alexander et al., 1993). The PR-1 proteins are developmentally regulated and expressed in response to external stimuli such as chemical elicitors (ethylene, SA, JA), wounding, hormones, and UV light (Buchel and Linthorst, 1999).

Chitinase Genes

Many of the genes and cDNAs encoding PR proteins have been isolated and characterized for several PR proteins over the past two decades. With the advancements in tools for genetic engineering, several PR-protein genes have been mobilized into plant systems and the resulting transgenic plants have been extensively studied. Most of the studies have resulted in the constitutive overexpression of PR proteins in all tissues, although some studies involve expression of these genes in the antisense orientation. However, the overexpression strategy incurs an additional metabolic load on the plant, diverting some of its resources to the synthesis of these defense proteins in an indiscriminate manner. Nevertheless, this strategy may be valuable if the goal is to enhance the level of resistance in susceptible germplasm and to stabilize the yield in geographic areas with high pathogen pressure (Datta et al., 1999). A more attractive approach would be to express combinations of PR-protein genes under the control of an inducible promoter or tissue-specific promoters.

Among the PR proteins, chitinases and β-1,3-glucanases are the best candidates for manipulating single-gene defense mechanisms because they target chitin and β-1,3-glucans, which are structural components of many fungi. Overexpression of a bean vacuolar chitinase resulted in significantly reduced symptoms of blight caused by *Rhizoctonia solani* in tobacco and canola plants (Broglie et al., 1991). This was the first successful report of the development of transgenic plants with a PR-protein gene showing enhanced levels of disease resistance. The constitutive expression of a class I rice chitinase gene, *chi11,* in rice delayed the progression of sheath blight and reduced lesion size when challenged with *R. solani* (Lin et al., 1995). Transgenic rice plants expressing another rice chitinase gene (*RC*-7) also had higher levels of resistance to sheath blight (Datta et al., 2001). A class I chitinase gene from barley was also effective in significantly reducing the disease symptoms due to sheath blight disease in transgenic rice when expressed at high levels (Ghareyazie et al., 2000).

Chitinase genes have also been introduced into other hosts in an attempt to enhance disease resistance. Only a limited number of studies involving some crop plants are reviewed here. Zhu and colleagues (1998) reported the introduction of a rice chitinase gene *(chi11)* into grain sorghum, under the control of the cauliflower mosaic virus 35S promoter *(CaMV 35S).* When homozygous progenies with stable high-level expression of the transgene were bioassayed against stalk rot, improved resistance to this disease was observed (Krishnaveni et al., 2001). Transgenic cucumber plants overexpressing a rice chitinase also exhibited enhanced resistance to gray mold disease caused by *Botrytis cinerea* (Tabei et al., 1998). Transgenic tobacco plants expressing a baculovirus-derived chitinase had higher levels of resistance to brown spot disease caused by *Alternaria alternata* (Shi et al., 2000). Similarly, a fungal chitinase gene from *Rhizopus oligosporus* was also shown to have antifungal activity in transgenic tobacco (Terakawa et al., 1997).

However, there are also studies in which there was no enhanced resistance in transgenic plants overexpressing particular chitinases. Transgenic tobacco plants overexpressing either a class I tobacco chitinase or an acidic class III chitinase from sugar beet did not exhibit increased resistance to *Cercospora nicotianae* (Neuhas et al., 1991; Nielsen et al., 1993). Similarly, we observed no reduction in disease symptoms due to *Fusarium* in transgenic wheat lines overexpressing a PR-3 class IV chitinase (unpublished data). Thus, chitinases may not be effective universally against all chitin-containing fungi. Many factors, such as the mode of infection by the pathogen, the nature of the specific chitinases, and their accessibility and location of chitin in the cell wall may have contributed to the success or failure of the attempts to enhance disease resistance.

β-1,3-Glucanase

Tobacco plants overexpressing a β-1,3-glucanase gene from soybean exhibited enhanced resistance to *Alternaria alternata* and *Phytophthora parasitica* var. *nicotianae* (Yoshikawa et al., 1993). Although β-1,3-glucanases appear to be tailored for defense against fungi, recent findings on mutants deficient for these enzymes generated with antisense transformation suggested their role in viral pathogenesis. Antisense β-1,3-glucanase transgenic tobacco that had lower levels of β-1,3-glucanase were less susceptible to TMV, suggesting that the decreased susceptibility to virus resulted from increased callose deposition in and around the lesions induced by TMV (Beffa et al., 1996). This also suggests the intriguing possibility that viruses could exploit the host defense mechanisms against other pathogens, e.g., fungi, to promote their own replication and spread.

TLP Gene (PR-5)

Transgenic tobacco plants overexpressing a tobacco osmotin gene did not display enhanced resistance to *Phytophthora parasitica,* but transgenic potato plants overexpressing the same osmotin gene were resistant to *P. infestans,* which causes late blight of potato (Liu et al., 1994). Transgenic potato plants overexpressing a native potato osmotin-like protein gene were also resistant to *P. infestans* (Zhu et al.,

1996). A TLP gene isolated from a rice cDNA library was introduced into the rice varieties Chinsura Boro II, IR 72, and IR1500 by protoplast mediated/biolistic transformation. Several independent transformants showed accumulation of the 23 kDa TLP at up to 0.5 percent of the total protein. When these high-expressing lines were challenged with the sheath blight pathogen *(Rhizoctonia solani),* disease symptoms were significantly reduced when compared to nonexpressing/nontransgenic plants (Velazhahan et al., 1998).

Oxalate Oxidase and **LTP** *Genes*

The ectopic expression of a cDNA clone for a wheat oxalate oxidase gene in hybrid poplar leaves increased its resistance to *Septoria musiva* (Liang et al., 2001). Transformation of tobacco and *Arabidopsis* with the gene for barley *LTP2* reduced necrotic lesions and decreased symptoms due to *Pseudomonas syringae* (Molina and Garcia-Olmedo, 1997).

Defensins

The expression of an alfalfa antifungal peptide (alfAFP) defensin isolated from seeds showed robust resistance to *Verticillium dahliae* in transgenic potato under greenhouse and field conditions (Gao et al., 2000). Similarly, the expression of a defensin gene *(WTI)* from *Wasabia japonica* using a potato virus X vector was demonstrated in tobacco. The defensin protein (WTI) could be purified from virus-infected transgenic leaves and had strong antifungal activity toward the phytopathogenic fungi *Magnaporthe griseae* and *Botrytis cinerea,* but reacted weakly against the bacterium *Pseudomonas cichorii* (Saitoh et al., 2001).

TRANSGENIC PLANTS WITH COMBINATIONS OF PR PROTEINS

Tomato plants individually expressing a chitinase/β-1,3-glucanase did not show any resistance to the *Fusarium* wilt pathogen, whereas combined expression of both of the transgenes achieved a higher level of resistance (Jach et al., 1995; Jongedijk et al., 1995). Tobacco lines expressing either a rice chitinase or an alfalfa glucanase were crossed to obtain hybrids containing both transgenes. The progenies were selfed to obtain homozygous plants which were then tested for resistance to frog eye spot. Transgenic plants expressing both genes showed greater resistance to disease when compared to single-gene controls, suggesting that chitinase and glucanase in combination was effective against the pathogen (Zhu et al., 1994).

Transgenic tobacco plants expressing a barley ribosome inactivating protein cDNA, under control of a wound-inducible promoter, were significantly protected against the soil-borne pathogen *R. solani* when compared to nontransgenic controls (Logemann et al., 1991). Higher levels of resistance to *R. solani* infection were observed in tobacco plants expressing three barley proteins, namely a class II chitinase, a β-1,3-glucanase, and a type I RIP. This observation supports the synergistic interaction of the PR proteins with other antifungal proteins, leading to enhanced levels of plant protection (Jach et al., 1995)

There are reports demonstrating the role of PR proteins in defense against plant pathogens outside the laboratory. These reports are, however, limited to greenhouse trials, and it would be interesting to show the feasibility and effect of these genes alone or in combination with others under field conditions.

Transformation of Wheat with Combinations of PR-Protein Genes

We have successfully transformed wheat (variety Bobwhite) with two different plasmids containing a total of four genes (two selectable markers, a rice chitinase gene *chi11*, and a rice *tlp* gene). The two genes under control of the maize ubiquitin *(ubi1)* promoter *(bar* and *tlp-D34* genes) were expressed at high levels even in the T_3 generation, whereas the other two genes *(hpt* and *chi11)* under control of *CaMV* 35S promoter were completely silenced in T_1 and later generations of all of the 18 independent transgenic events studied (Chen et al., 1998, 1999). These results suggested that the 35S promoter was susceptible to gene silencing, while the *ubi1* promoter appeared to be stably inherited and functioned normally. The transgenic wheat plants with the *tlp* gene accumulating the transgenic protein at high levels were tested against the scab fungus *Fusarium graminearum* and were found to be significantly less susceptible than control non-transgenic plants in greenhouse trials (Chen et al., 1999). The control plants showed infection in 43 percent of the spikes, whereas the transgenic plants had infection in only 16 percent of the spikes, based on observations made ten days after inoculation. However, this difference became less significant 14 days after inoculation. Although the rice *tlp* gene could delay the progression of scab disease, it could not offer complete resistance (Chen et al., 1999).

In an effort to identify and characterize candidate PR-protein genes that might be effective against scab, a scab-inoculated cDNA library of Sumai-3, a Chinese spring variety with type II resistance, was screened (Li et al., 2001). Two inducible acidic chitinases (PR-3 class IV and VII) and two different acidic glucanases were identified from this library. These genes were subsequently placed under the control of a constitutive promoter, maize ubiquitin promoter-intron *(ubi1)*, and used for cotransformation of wheat with the selectable marker *bar* gene (for herbicide resistance) under the same *ubi1* promoter. Twenty-six independent T_0 transgenic wheat lines were identified that were resistant to herbicide and tested positive for the transgene and the transgene-encoded protein (Anand et al., 2000, 2001). A high cointegration frequency (90 percent) of transgenes was observed in cotransformation experiments with the two plasmids, demonstrating the feasibility of cotransformation for obtaining transgenic wheat plants, making the use of large cointegrate vectors unnecessary. We identified five transgenic wheat lines with stable inheritance and expression of different combinations of the transgenes in subsequent generations. One particular line expressing high levels of class IV chitinase *(chi)* and a glucanase *(glu)* gene combination showed decreased scab *(F. graminearum)* symptoms when compared with the nontransgenic control (Anand et al., unpublished data). Most of the transgene-encoded and endogenous PR proteins were localized in the apoplastic fluids. The other four lines stably expressing either a single gene *(chi* or *glu)* or a combination of two genes were propagated to the T_4 generation. These lines had lower levels of transgene-encoded proteins and did not show any increased resistance to scab (unpublished data). We are attempting to determine whether a threshold level or a specific combination of PR proteins is required for effective resistance to scab.

Transgenic lines of wheat expressing three different antifungal genes individually (an *afp* gene from the fungus *Aspergillus giganteus,* a barley class II chitinase gene and a barley type I RIP gene under the maize ubiquitin promoter) stably inherited and expressed the transgenes over four generations. The expression of the fungal *afp* or barley seed-specific chitinase gene in wheat resulted in enhanced resistance to powdery mildew and leaf rust, whereas RIP did not confer any resistance (Oldach et al., 2001). Transient expression of two different representatives of cereal germinlike proteins (*gf-2.8* and *TaGLP2a*) in wheat conferred penetration resistance to wheat epidermal cells that were attacked by *Blumeria graminis* f. sp. *tritici,* the powdery mildew fungus (Schweizer et al., 1999).

The gene-pyramiding approach may prove to be an effective way of enhancing disease resistance of the wheat germplasm to various wheat diseases, because different cellular components of the pathogen are the targets for PR proteins. In this context, we have already crossed transgenic lines expressing different combinations of PR-protein genes. For example, we have advanced lines containing the *tlp/ch/glu* genes. The homozygous transgenic wheat lines with all three genes will be tested in the greenhouse and in the field to evaluate differences in levels of scab resistance.

Gene Silencing in Transgenic Plants

Introduction of transgenes into plants often results in decreased expression of both the transgene and its resident homologue. Constitutive expression of a vacuolar targeted chimeric class I tobacco chitinase gene from tobacco in *Nicotiana sylvestris* under regulation of the *CaMV* 35S promoter resulted in higher levels of accumulation of chitinase in transgenic plants homozygous for the transgene locus. However, some of the T_2 progenies had lower levels of both endogenous and transgene-coded chitinase mRNAs (Hart et al., 1992). Some direct descendants of homozygous lines had a high-low-high pattern of expression in subsequent generations. These results suggest that such silent phenotypes result from stable but potentially reversible inactivation of genes that cause cosuppression of both the host and transgenes.

De Carvalho and colleagues (1992) observed cosuppression of a β-1,3-glucanase gene in homozygous transgenic plants, whereas in the heterozygous plants, the transgene-encoded protein was present at very high levels when compared to nontransgenic controls. Cosuppression occurred posttranscriptionally and was developmentally controlled. This phenomenon was directly correlated with the dose of transgene in the plant genome.

Our studies on the expression of a rice chitinase gene *(chi11)* under control of the *CaMV* 35S promoter in transgenic rice revealed that all T_3 progeny plants homozygous for the transgene locus were expressing the gene constitutively for up to three weeks after germination. However, 20 percent of these progeny did not show any expression at detectable levels eight weeks after germination, indicating a silencing of the transgene had occurred in a proportion of the T_3 plants. There was cosilencing of the selectable marker gene, *bar,* which was also under control of the *CaMV* 35S promoter. The silenced phenotype was not reversed in the next generation, indicating a permanent alteration, even though no obvious alteration in the transgene could be demonstrated. Northern blot and nuclear run-on transcriptional analyses revealed that the metabolic block in the silent plants was transcriptional (Chareonpornwattana et al., 1999). We also noticed silencing of rice chitinase in a transgenic sorghum line to a similar extent (Zhu et al., 1998).

Transgenic wheat cobombarded with two different plasmids with a total of four genes (two selectable markers *hpt* and *bar,* a rice chitinase *chi11,* and a rice *tlp* gene) were studied to understand the mechanism of silencing. The *chi11* gene under the *CaMV* 35S promoter expressed the transgene-encoded protein in T_0 plants, while the T_1 plants did not show any expression of the transgene. The *CaMV* 35S promoter was inactivated more rapidly in wheat than in rice. Almost none of the T_1 progeny had detectable levels of chitinase (Chen et al., 1998, 1999). Even the selectable marker gene *(hpt)* under the control of another *CaMV* 35S promoter was silenced. However, two other genes in the same transgenic plant and their progenies which were under the control of the maize *ubi1* promoter expressed the transgenes (a rice *tlp* and *bar* gene) even in the T_3 generation (Chen et al., 1999). It was concluded that the *CaMV* 35S promoter was not a suitable promoter for expression of transgenes in wheat.

Thus, a change in the strategy for wheat transformation using the maize *ubi1* promoter was employed. However, more recent studies in our laboratory have revealed that a majority of the transgenic wheat lines (80 percent) with the *bar* gene and the gene of interest (both driven by *ubi1* promoter intron) were completely silenced in T_1/T_2 generation, demonstrating that the *ubi1* promoter was also susceptible to random and frequent gene silencing. In an interesting case, two plants cotransformed with a combination of wheat *chi* and *glu* transgenes had identical patterns of transgene bands in Southern blot analyses but had totally different patterns of expression of the transgenes. One line had stable, high-level expression of both transgenes at the RNA and protein levels over four generations, while the other line was completely silenced in the T_3 generation and beyond. Southern blot analyses of genomic DNA from these lines using the methylation-sensitive enzymes *MspI* and *HpaII* (isoschizomers recognizing CCGG and its methylated forms) revealed differences in the degree of methylation of the transgene locus between these lines. We have attributed transgene methylation as a possible reason for the observed differences in expression and gene silencing (Anand et al., unpublished data). Independent wheat transgenic lines differing in copy number for the transgene (3 to 15 copies) were also identified among 24 transgenic lines that we have analyzed and we could not find any correlation between transgene expression level and copy number. It appears that gene silencing in wheat was more a random phenomenon that was not dependent on the copy number, the choice of promoter, or the site of integration.

CONCLUSIONS

The wide distribution of pathogen-inducible genes in the plant kingdom suggests that the proteins encoded by these genes contribute significantly to the resistance of plants against parasitic attack. When pathogens invade host tissues, these genes are rapidly activated and a number of specific PR proteins are synthesized. Many of these proteins have antifungal and antibacterial or insecticidal activity. Among the PR proteins, chitinases and glucanases are the most commonly studied because of their strong known in vitro antifungal activity and known cell wall–associated targets. Apart from their antifungal activity, chitinases and glucanases are also believed to be involved in some other plant functions. PR proteins might also be involved in pathogen recognition processes, by binding to pathogen cell wall components and/or by generating signal molecules from pathogens. Certain PR proteins are involved in stress tolerance and may play a role in plant morphogenesis. Most PR proteins are synthesized on polysomes that are membrane bound, and once synthesized, they are secreted into the apoplast and/or vacuoles. The specific targeting domains or transport competence determines the specificity of the site of accumulation. The acidic forms of PR proteins predominantly accumulate in the apoplast, while basic ones accumulate in the vacuoles of host cells, although there are notable exceptions to this rule. Among the PR proteins, basic ones have been shown to have strong antifungal activity. The acidic PR proteins are likely to be involved in signal generation. Various signal molecules produced by pathogenic or nonpathogenic organisms or derived from plant-pathogen interactions can induce PR-protein genes. In addition, a variety of chemical signal molecules, such as SA, JA, ethylene, and β-amino *n*-butanoic acid can induce PR-protein genes. Although PR proteins are likely to be directly and indirectly involved during the infection process, pathogens have coevolved and may have developed mechanisms that can protect them from the effects of some of these PR proteins. This is one of the criteria that should be considered when expressing any PR-protein gene in plants to improve disease resistance.

During the past decade, many researchers have overexpressed various PR-protein genes in plants and studied their function and contribution to disease resistance. Although there are abundant reports on the successful expression of PR proteins in plants and improved resistance to pathogens under controlled conditions, the paucity of field data with regard to sustainable levels of resistance to diseases over many generations has cast some doubt about the utility of these genes. It is likely that in nature, where plants are exposed to different pathogens and insects within the same generation and over several generations, a broad-spectrum resistance is more valuable than total resistance to one pathogen (typified by the gene-for-gene interaction involving *avr/R* gene interactions). The strategy involving multiple PR proteins may not be completely effective against a single race of the pathogen but will be useful in achieving broad-spectrum resistance and in slowing down the development of resistant races of the pathogens and pests.

Additional research aimed at understanding the biochemical basis of antimicrobial and insecticidal activity of these PR proteins in plant defense is needed. For instance, except for a few groups such as PR-2, PR-3, and PR-6, the precise biochemical role of several of the PR proteins is either not known or incompletely understood. Even though PR-1 is the predominant PR protein that accumulates to a level of 1 to 2 percent of total proteins in infected plants, little is known about the exact function of this protein. The utility of these PR-protein genes will be fully realized only when we understand the targets and modes of action of ubiquitous PR proteins.

REFERENCES

Abad, L.R., D'Urzo, M.P., Liu, D., Narasimhan, M.L., Reuveni, M., Zhu, J.K., Niu, X., Singh, N.K., Hasegawa, P.M., and Bressan, R.A. (1996). Antifungal activity of tobacco osmotin has specificity and involves plasma membrane permeabilization. *Plant Science* 118: 11-23.

Ahl, P. and Gianinazzi, S. (1982). b-Protein as a constitutive component in highly resistant interspecific hybrids of *Nicotiana glutinosa* × *Nicotiana debneyi*. *Plant Science Letters* 26: 173-182.

Alexander, D., Goodman, R.M., Gut-Rella, M., Glascock, C., Weymann, K., Friedrich, L., Maddox, D., Ahl-Goy, P., Luntz, T., Ward, E., and Ryals, J. (1993). Increased tolerance to two oomycete pathogens in transgenic tobacco expressing pathogenesis-related protein 1a. *Proceedings of the National Academy of Sciences, USA* 90:7327-7331.

Anand, A., Janakiraman, V., Zhou, T., Trick, H.N., Gill, B.S., and Muthukrishnan, S. (2001). Transgenic wheat overexpressing PR-proteins shows a delay in *Fusarium* head blight infection. In *Proceedings of the National Fusarium Head Blight Forum* (pp. 2-6). Cincinnati, OH: U.S. Wheat and Barley Scab Initiative.

Anand, A., Li, W., Sakthivel, N., Krishnaveni, S., Muthukrishnan, S., Gill, B.S., Essig, J.S., Adams, R.E., Janakiraman, V., and Trick, H.N. (2000). Characterization of wheat PR-proteins cDNA's for transformation of wheat to enhance resistance to scab. In *Proceedings of the National Fusarium Head Blight Forum* (pp. 5-12). Cincinnati, OH: U.S. Wheat and Barley Scab Initiative.

Anguelova-Merhar, V.S., Van Der Westhuizen, and Pretorius, Z.A. (2001). β-1,3-glucanase and chitinase activities and the response of wheat to leaf rust. *Journal of Phytopathology* 149: 381-384.

Antoniw, J.F., Ritter, C.E., Pierpoint, W.S., and Van Loon, L.C. (1980). Comparison of the pathogenesis-related proteins from plants of two cultivars of tobacco infected with TMV. *Journal of General Virology* 47: 79-87.

Beffa, R.S., Hofer, R.M., Thomas, M., and Meins, F., Jr. (1996). Decreased susceptibility to viral disease of β-1,3-glucanase-deficient plants generated by antisense transformation. *Plant Cell* 8: 1001-1011.

Belles, J.M., Tomero, O., and Conejero, V. (1992). Pathogenesis-related proteins and polyamines in a developmental mutant of tomato Epinastic. *Plant Physiology* 98: 1502-1505.

Blilou, I., Juan, A., Ocampo, J.A., and Carcia-Garrido, J.M. (2000). Induction of LTP (lipid transfer protein) and PAL (phenylalanine ammonia-lyase) gene expression in rice roots colonized by the arbuscular mycorrhizal fungus *Glomus mosseae*. *Journal of Experimental Botany* 51: 1969-1977.

Bohlmann, H. (1999). The role of thionins in the resistance of plants. In Datta, S.K. and Muthukrishnan, S. (Eds.), *Pathogenesis-Related Proteins in Plants* (pp. 207-234). New York: CRC Press.

Broekaert, W.F., Cammue, B.P.A., and De Bolle, M.F.C. (1997). Antimicrobial peptides from plants. *Critical Reviews in Plant Science* 16: 297-323.

Broglie, W.F., Chet, I., Holliday, M., Cressman, R., Biddle, P., Knowlton, S., Mauvais, C.J., and Broglie, R. (1991). Transgenic plants with enhanced resistance to the fungal pathogen *Rhizoctonia solani*. *Science* 254: 1194-1197.

Buchel, A.S. and Linthorst, H.J.M. (1999). PR-1: A group of plant proteins induced upon pathogen infection. In Datta, S.K. and Muthukrishnan, S. (Eds.), *Pathogenesis-Related Proteins in Plants* (pp. 21-48). New York: CRC Press.

Carmona, M.J., Molina, A., and Fernandez, J.A. (1993). Expression of the alpha-thionin gene from barley in tobacco confers enhanced resistance to bacterial pathogens. *Plant Journal* 3: 457-462.

Chareonpornwattana, S., Thara, K.V., Wang, L., Datta, S.K., Panbangred, W., and Muthukrishnan, S. (1999). Inheritance, expression and silencing of a chitinase transgene in rice. *Theoretical and Applied Genetics* 98: 371-378.

Chen, W.P., Chen, P.D., Liu, D.J., Kynast, R.J., Friebe, B., Velazhahan, R., Muthukrishnan, S., and Gill, B.S. (1999). Development of wheat scab symptoms is delayed in transgenic wheat plants that constitutively express a rice thaumatin-like protein gene. *Theoretical and Applied Genetics* 99: 755-760.

Chen, W.P., Gu, X., Liang, G.H., Muthukrishnan, S., Chen, P.D., Liu, D.J., and Gill, B.S. (1998). Introduction and constitutive expression of a rice chitinase gene in bread wheat using biolistic bombardment and the *bar* gene as a selectable marker. *Theoretical and Applied Genetics* 97: 1296-1306.

Chittoor, M. J., Leach, J.E. and White, F.F. (1999). Induction of peroxidase during defense against pathogens. In Datta, S.K. and Muthukrishnan, S. (Eds.), *Pathogenesis-Related Proteins in Plants* (pp. 171-194). New York: CRC Press.

Conceição, A.S. and Broekaert, W.F. (1999). Plant defensins. In Datta, S.K. and Muthukrishnan, S. (Eds.), *Pathogenesis-Related Proteins in Plants* (pp. 247-260). New York: CRC Press.

Cornelissen, B.J.C., Hooft van Huijduijnen, R.A.M., and Bol, J.F. (1986). A tobacco mosaic virus-induced tobacco protein is homologous to the sweet-tasting protein thaumatin. *Nature* 321: 531-532.

Coventry, H.S. and Dubery, I.A. (2001). Lipopolysaccharides from *Burkholderia cepacia* contribute to an enhanced defensive capacity and the induction of pathogenesis-related proteins in *Nicotiana tabacum. Physiological and Molecular Plant Pathology* 58: 149-158.

Datta, K., Jumin, T., Oliva, N., Ona, I., Velazhahan, R., Mew, T.W., Muthukrishnan, S., and Datta, S.K. (2001). Enhanced resistance to sheath blight by constitutive expression of infection-related rice chitinase in transgenic elite indica rice cultivars. *Plant Science* 160: 405-414.

Datta, K., Muthukrishnan, S., and Datta, S. (1999). Expression and function of PR-protein genes in transgenic plants. In Datta, S.K. and Muthukrishnan, S. (Eds.), *Pathogenesis-Related Proteins in Plants* (pp. 261-278). New York: CRC Press.

De Carvalho, F., Gheysen, G., Kushnir, S., Van Montagu, M., Inze, D., and Catresana, C. (1992). Suppression of β-1,3-glucanase transgene expression in homozygous plants. *EMBO Journal* 11: 2595-2602.

Dixelius, C. (1994). Presence of the pathogenesis-related proteins 2, Q and S in stressed *Brassica napus* and *B. nigra* plantlets. *Physiological and Molecular Plant Pathology* 44:1-8.

Dixon, D.C., Cutt, J.R., and Klessig, D.F. (1991). Differential targeting of the tobacco PR-1 pathogenesis-related proteins to the extracellular space and vacuoles of crystal idioblasts. *EMBO Journal* 10: 1317-1324.

Epple, P., Apel, K., and Bohlmann, H. (1997). ESTs reveal a multigene family for plant defensins in *Arabidopsis thaliana. FEBS Letters* 400: 168-172.

Ernst, D., Schraudner, M., Langebartels, C., and Sandermann, H., Jr. (1992). Ozone-induced changes of mRNA levels of β-1,3-glucanase, chitinase and pathogenesis-related protein 1b in tobacco plants. *Plant Molecular Biology* 20: 673-682.

Eyal, Y., Sagee, O., and Fluhr, R. (1992). Dark induced accumulation of a basic pathogenesis-related (PR-1) transcript and a light requirement for its induction by ethylene. *Plant Molecular Biology* 19: 589-599.

Farmer, E.E., Johnson, R.R., and Ryan, C.A. (1992). Regulation of expression of proteinase inhibitor genes by methyl jasmonate and jasmonic acid. *Plant Physiology* 98: 995-1002.

Felix, G. and Meins, F. (1987). Ethylene regulation of β-1,3-glucanase in tobacco. *Planta* 172: 386-392.

Gaffney, T., Friedrich, L., Vernooij, B., Negrotto, D., Nye, G., Uknes, S., Ward, E., Kessmann, H., and Ryals, J. (1993). Requirement of salicylic acid for the induction of systemic acquired resistance. *Science* 261: 754-756.

Gao, A.G., Hakimi, S.M., Mittanck, C.A., Wu, Y., Woerner, B.M., Stark, D.M., Shah, D.M, Liang, J., and Rommens, C.M. (2000). Fungal pathogen protection in potato by expression of a plant defensin peptide. *Nature Biotechnology* 18: 1307-1310.

Garcia-Olmedo, F., Molina, A., Segura, A., and Moreno, M. (1995). The defensive role of nonspecific lipid-transfer proteins in plants. *Trends in Microbiology* 3: 72-74.

Garcia-Olmedo, F., Rodriguez-Palenzuela, P., Hernandez-Lucas C., Ponz F., Marana, C., Carmona, M.J., Lopez-Fando J.J., Fernandez, J.A., and Carbonero, P. (1989). The thionins: A protein family that includes purothionins, viscotoxins and crambin. In Miflin, B.J. (Ed.), *Oxford Surveys of Plant Molecular and Cell Biology* 6 (pp. 31-60). Oxford, UK: Oxford University Press.

Ghareyazie, B., Menguito, C., Rubia, L.G., De Palma, J., Ona, A., Muthukrishnan, S., Velazhahan, R., Kush, G.S., and Bennet, J. (2000). Insect resistant aromatic rice is expressing a barley chitinase gene and is resistant to sheath blight. In Peng, S. and Hardy, H. (Eds.), *Proceedings of the International Rice Research Conference* (pp. 692-701). Las Banos, Philippines: International Rice Research Institute.

Gordon-Weeks, R., White, R.F., and Pierpoint, W.S. (1991). Evidence for the presence of endogenous inducers of the accumulation of pathogenesis-related (PR-1) proteins in leaves of virus-infected tobacco plants and of an interspecific *Nicotiana* hybrid. *Physiological and Molecular Plant Pathology* 38: 375-392.

Hae-Keun, Y., So-Young, Y., Seung-Heun, Y., and Doil, C. (1999). Cloning of pathogenesis-related protein-I gene from *Nicotiana glutinosa* and its salicylic acid-independent induction by copper and β-aminobutyric acid. *Journal of Plant Physiology* 154: 327-333.

Hart, C.M., Fischer, B., Neuhas, J.M., and Meins, F., Jr. (1992). Regulated inactivation of homologous gene expression in transgenic *Nicotiana sylvestris* plants containing a defense-related tobacco chitinase gene. *Molecular General Genetics* 235: 179-184.

Hass, G.M., Hermodson, M.A., and Ryan, C.A. (1982). Primary structures of two low molecular weight proteinase inhibitors from potatoes. *Biochemistry* 21: 2-9.

Heitz, T., Geoffroy, P., Fritig, B., and Legrand, M. (1999). The PR-6 family: Proteinase inhibitors in plant-microbe and plant-insect interactions. In Datta, S.K. and Muthukrishnan, S. (Eds.), *Pathogenesis-Related Proteins in Plants* (pp. 131-155). New York: CRC Press.

Hilder, V.A., Gatehouse, D.M.R., Sheerman, S.E., Barker, R.F., and Boulter, D. (1987). A novel mechanism of insect resistance engineered into tobacco. *Nature* 330: 160-163.

Hurkman, W.J. and Tanaka, C.K. (1996) Germin gene expression is induced in wheat leaves by powdery mildew infection. *Plant Physiology* 111: 735-739.

Ishige, F., Mori, H., Yamazaki, K., and Imaseki, H. (1991). The effect of ethylene on the coordinated synthesis of multiple proteins: Accumulation of an acidic chitinase and a basic glycoprotein induced by ethylene in leaves of azuki beans, *Vigna angulasis. Plant Cell Physiology* 32: 681-690.

Jach, G., Gornhardt, B., and Mundy, J. (1995). Enhanced quantitative resistance against fungal disease by combinatorial expression of different barley antifungal proteins in transgenic tobacco. *Plant Journal* 8: 97-109.

Jacob, G., Cottier, V., Toquin, V., Rigoli, G., Zimmerli, L., Metraux, J.P., and Mauch-Mani, B. (2001). β-Aminobutyric acid-induced resistance in plants. *European Journal of Plant Pathology* 107: 29-37.

Jongedijk, E., Tigelaar, H., and Van Roekel, J.S.C. (1995). Synergetic activity of chitinases and β-1,3-glucanases enhances fungal resistance in transgenic tomato plants. *Euphytica* 85: 173-180.

Kader, J.C. (1996). Lipid transfer proteins in plants. *Annual Review of Plant Physiology and Plant Molecular Biology* 47: 627-654.

Kim, Y.S., Yoon, K., Liu, J.R., and Lee, H.S. (1997). Characterizaton of CHRK1, a receptor-like kinase containing a chitinase-related domain in its N-terminus. Abstracts of the Fifth International Congress of Plant Molecular Biology, Singapore, September 21-27 (p. 789).

Kragh, K.M., Hendriks, T., De Jong, A.J., Lo Schiavo, F., Bucherna, N., Hojrup, P., Mikkelsen, J.D., and De Vries, S.C. (1996). Characterization of chitinases able to rescue somatic embryos of the temperature-sensitive carrot variant *ts11*. *Plant Molecular Biology* 31: 631-635.

Krebs, S.L. and Grumet, R. (1993). Characterization of celery hydrolytic enzymes produced in response to infection by *Fusarium oxysporum*. *Physiological and Molecular Plant Pathology* 43: 193-208.

Krishnaveni, S., Jeoung, J.M., Muthukrishnan, S., and Liang, G.H. (2001). Transgenic sorghum plants constitutively expressing a rice chitinase gene show improved resistance to stalk rot. *Journal of Genetics and Breeding* 55: 151-158.

Krishnaveni, S., Muthukrishnan, S., Liang, G.H., Wilde, G., and Manickam, A. (1999). Induction of chitinases and β-1,3-glucanases in resistant and sensitive cultivars of sorghum in response to insect attack, fungal infection and wounding. *Plant Science* 144: 9-16.

Kuc, J. (2001). Concepts and direction of induced systemic resistance in plants and its application. *European Journal of Plant Pathology* 107: 7-12.

Lane, B.G. (1994). Oxalate, germin and the extracellular matrix of higher plants. *FASEB Journal* 8: 294-301.

Lawton, K.A., Potter, S.L., Uknes, S., and Ryals, J. (1994). Acquired resistance: Signal transduction in *Arabidopsis* is ethylene independent. *Plant Cell* 6: 581-588.

Leubner-Metzger, G. and Meins, F., Jr. (1999). Functions and regulation of plant β-1,3-glucanase (PR-2). In Datta, S.K. and Muthukrishnan, S. (Eds.), *Pathogenesis-Related Proteins in Plants* (pp. 49-76). New York: CRC Press.

Li, W.L., Faris, J.D., Muthukrishnan, S., Liu, D.J., Chen, P.D., and Gill, B.S. (2001). Isolation and characterization of novel cDNA clones of acidic chitinases and β-1,3-glucanases from wheat spikes infected with *Fusarium graminearum*. *Theoretical and Applied Genetics* 102: 353-362.

Liang, H., Maynard, C.A., Allen, R.A., and Powell, W.A. (2001). Increased *Septoria musiva* resistance in transgenic hybrid poplar leaves expressing a wheat oxalate oxidase gene. *Plant Molecular Biology* 45: 619-629.

Lin, W., Anuratha, C.S., Datta, K., Potrykus, I., Muthukrishnan, S., and Datta, S.K. (1995). Genetic engineering of rice for resistance to sheath blight. *Bio/Technology* 13: 686-691.

Linthorst, H.J.M., Meuwissen, R.L.J., Kauffmann, S., and Bol, J.F. (1989). Constitutive expression of pathogenesis-related proteins PR-1, GRP, and PR-S in tobacco has no effect on virus infection. *Plant Cell* 1: 285-291.

Liu, D., Ragothama, K.G., Hasegawa, P.M., and Bressan, R.A. (1994). Osmotin over-expression in potato delays development of disease symptoms. *Proceedings of the National Academy of Sciences, USA* 91: 1888-1892.

Logemann, J., Jach., G., and Tommerup, H. (1991). Expression of a barley ribosome-inactivating protein leads to increased fungal protection in transgenic tobacco plants. *Bio/Technology* 10: 305-308.

Malamy, J., Carr, J.P., Klessig, D.F., and Raskin, I. (1990). Salicylic acid: A likely endogenous signal in the resistance response of tobacco to viral infection. *Science* 250: 1002-1004.

Matsuoka, K., Matsumoto, S., Hattori, T., Machida, Y., and Nakamura, K. (1990). Vacuolar targeting and posttranslational processing of the precursor to the sweet potato tuberous root storage protein in heterologous plant cells. *Journal of Biological Chemistry* 265: 19750-19757.

Matsuoka, M. and Ohashi, Y. (1986). Induction of pathogenesis-related proteins in tobacco leaves. *Plant Physiology* 80: 505-510.

Mauch, F., Hadwiger, L.A., and Boller, T. (1984). Ethylene: Symptom not signal for the induction of chitinase and β-1,3-glucanase in pea pods by pathogens and elicitors. *Plant Physiology* 76: 607-611.

Mauch, F., Mauch-Mani, B., and Boller, T. (1988). Antifungal hydrolases in pea tissue: II. Inhibition of fungal growth by combinations of chitinase and β-1,3-glucanase. *Plant Physiology* 88: 936-942.

McGurl, N., Pearce, G., Orozco-Cardenas, M., and Ryan, C.A. (1992). Structure, expression and antisense inhibition of the systemin precursor gene. *Science* 255: 1570-1573.

Memelink, J., Linthorst, H.J.M., and Schilperoort, R.A. (1990). Tobacco genes encoding acidic and basic isoforms of pathogenesis-related proteins display different expression patterns. *Plant Molecular Biology* 14: 119-126.

Metraux, J.P. (2001). Systemic acquired resistance and salicylic acid: Current state of knowledge. *European Journal of Plant Pathology* 107: 13-18.

Molina, A. and Garcia-Olmedo, F. (1997). Enhanced tolerance to bacterial pathogens caused by the transgenic expression of barley lipid transfer protein LTP2. *Plant Journal* 12: 669-675.

Muthukrishnan, S., Liang, G.H., Trick, H.N., and Gill, B.S. (2001). Pathogenesis-related proteins and their genes in cereals. *Plant Cell Tissue Organ Culture* 64: 93-114.

Neuhaus, J.M. (1999). Plant chitinases (PR-3, PR-4, PR-4, PR-8, PR-11). In Datta, S.K. and Muthukrishnan, S. (Eds.), *Pathogenesis-Related Proteins in Plants* (pp. 77-106). New York: CRC Press.

Neuhaus, J.M., Ahl-Goy, P., and Hinz, U. (1991). High-level expression of a tobacco chitinase gene in *Nicotiana sylvestris:* Susceptibility of transgenic plants
to *Cercospora nicotianae* infection. *Plant Molecular Biology* 16: 141-151.

Newman, M.A., Dow, J.M., and Daniels, M.J. (2001). Bacterial lipopolysaccharides and plant-pathogen interactions. *European Journal of Plant Pathology*
107: 95-102.

Nielsen, K.K., Mikkelsen, J.D., Kragh, K.M., and Bojsen, K. (1993). An acidic
class III chitinase in sugar beet induction by transgenic tobacco plants. *Molecular Plant-Microbe Interactions* 6: 495-506.

O'Connell, R.J. and Ride, J.P. (1990). Chemical detection and ultrastructural localization of chitin in cell walls of *Colletotrichum lindemuthianum. Physiological
and Molecular Plant Pathology* 37: 39-53.

Oldach, K.H., Becker, D., and Lorz, H. (2001). Heterologous expression of genes
mediating enhanced fungal resistance in transgenic wheat. *Molecular Plant-
Microbe Interactions* 14: 832-838.

Oostendorp, M., Kunz, W., Deitrich, B., and Staub, T. (2001). Induced disease resistance by chemicals. *European Journal of Plant Pathology* 107: 19-28.

Pearce, G., Strydom, D., Johnson, S., and Ryan, C.A. (1991). A polypeptide from
tomato polyamines in a developmental mutant of tomato Epinastic. *Plant Physiology* 98: 1502-1505.

Peng, J.H. and Black, L.L. (1976). Increased protein inhibitor activity in response to
infection of resistant tomato plants by *Phytophthora infestans. Phytopathology*
66: 958-963.

Porat, R., Mccollum, T.G., Vinokur, V., and Droby, S. (2002). Effects of various
elicitors on the transcription of a β-1,3-endoglucanase gene in citrus fruit. *Journal of Phytopathology* 150: 70-76.

Punja, Z.K. (2001). Genetic engineering of plants to enhance resistance to fungal
pathogens—A review of progress and future prospects. *Canadian Journal of
Plant Pathology* 23: 216-235.

Pyee, J., Yu, H., and Kolattukudy, P.E. (1994). Identification of a lipid transfer protein as the major protein in the surface wax of broccoli *(Brassica oleracea)*
leaves. *Archives of Biochemistry and Biophysics* 311: 460-468.

Ramamoorthy, V., Viswanathan, R., Raguchander, T., Prakasam, V., and Sami-
yappan, R. (2001). Induction of systemic resistance by plant growth promoting
rhizobacteria in crop plants against pests and diseases. *Crop Protection* 20: 1-11.

Rameshsundar, A.R., Velazhahan, R., Viswanathan, R., Padmanaban, P., and
Vidhyasekaran, P. (2001). Induction of systemic signal molecule, acibenzolar-*S*-
methyl. *Phytoparasitica* 29: 231-242.

Reimmann, C. and Dudler, R. (1993). cDNA cloning and sequence analysis of a
pathogen-induced thaumatin-like protein from rice *(Oryza sativa). Plant Physiology* 101: 1113-1114.

Reiss, E. and Horstmann, C. (2001). *Drechslera teres*-infected barley *(Hordeum
vulgare* L.) leaves accumulate eight isoforms of thaumatin-like proteins. *Physiological and Molecular Plant Pathology* 58: 183-188.

Ride, J.P. (1983). Cell walls and other structural barriers in defense. In Callow, J.A. (Ed.), *Biochemical Plant Pathology* (pp. 215-236). New York: Wiley-Interscience.

Rodrigo, I., Vera, P., and Conejero, V. (1989). Degradation of tomato pathogenesis-related proteins by an endogenous aspartyl endoproteinase. *European Journal of Biochemistry* 184: 663-669.

Ryan, C.A. (1990). Protease inhibitors in plants: Genes for improving defenses against insects and pathogens. *Annual Review of Phytopathology* 28: 425-449.

Saitoh, H., Kiba, A., Nishihara, M., Yamamura, S., Suzuki, K., and Terauchi, R. (2001). Production of antimicrobial defensins in *Nicotiana benthamiana* with a potato virus X vector. *Molecular Plant-Microbe Interactions* 14: 111-115

Schweizer, P., Christoffel, A., and Dudler, R. (1999). Transient expression of members of the *germin*-like gene in epidermal cells of wheat confers disease resistance. *Plant Journal* 20: 541-552.

Sharp, J.K., Valent, B., and Albersheim, P. (1984). Purification and partial characterization of a β-glucan fragment that elicits phytoalexin accumulation in soybean. *Journal of Biological Chemistry* 259: 11312-11320.

Shi, J., Thomas, C.J., King, L.A., Hawes, C.R., Posee, R.D., Edwards, M.L., Pallet, D., and Cooper, J.I. (2000). The expression of baculovirus-derived chitinase gene increased resistance of tobacco cultivars to brown spot *(Alternaria alternata)*. *Annals of Applied Biology* 136: 1-8.

Seigrist, J. and Kauss, H. (1990). Chitin deacetylase in cucumber leaves infected by *Colletotrichum lagenarium*. *Physiological and Molecular Plant Pathology* 36: 267-275.

Siegrist, J., Orober, M., and Buchenauer, H. (2000). β-Aminobutyric acid-mediated enhancement of resistance in tobacco to tobacco mosaic virus depends on the accumulation of salicylic acid. *Physiological and Molecular Plant Pathology* 56: 95-106.

Silue, D., Pajot, E., and Cohen, Y. (2002). Induction of resistance to downy mildew *(Peronospora parasitica)* in cauliflower by DL-β-amino-n-butanoic acid (BABA). *Plant Pathology* 51: 97-102.

Staehelin, C., Granado, J., Muller, J., Wiemken. A., Mellor, R.B., Felix, G., Regenass, M., Broughton, W.J., and Boller, T. (1994). Perception of *Rhizobium* nodulation factors by tomato cells and inactivation by root. *Proceedings of the National Academy of Sciences, USA* 91: 2196-2200.

Sterk, P., Booij, H., Schellekens, G.A., Van Kammen, A., and De Vries, S.C. (1991). Cell-specific expression of the carrot EP2 lipid transfer protein gene. *Plant Cell* 3: 907-921.

Tabei, Y., Kitade, S., Nishizawa, Y., Kikuchi, N., Kayano, T., Hibi, T., and Akutsu, K. (1998). Transgenic cucumber plants harboring a rice chitinase gene exhibited enhance resistance to grey mould *(Botrytis cinerea)*. *Plant Cell Reports* 17: 159-164.

Terakawa, T., Takaya, N., Horiuchi, H., Koike, M., and Takagi, M. (1997). A fungal chitinase gene from *Rhizopus oligosporus* confers antifungal activity to transgenic tobacco. *Plant Cell Reports* 16: 439-442.

Thevissen, K., Osborn, R.W., Acland, D.P., and Broekaert, W.F. (1997). Specific, high affinity binding sites for an antifungal plant defensin on *Neurospora crassa* hyphae and microsomal membranes. *Journal of Biological Chemistry* 272: 32176-32181.

Van Loon, L.C. (1990). The nomenclature of pathogenesis-related proteins. *Physiological and Molecular Plant Pathology* 37: 229-230.

Van Loon, L.C. (2001). The families of pathogenesis-related proteins. In Proceedings of the Sixth International Workshop on PR-Proteins, Belgium, Spa, May 20-24 (pp. 7-10) .

Van Loon, L.C. and Van Kammen, A. (1970). Polyacrylamide disc electrophoresis of the soluble leaf proteins from *Nicotiana tabacum* var. 'Samsun' and 'Samsun NN'. II. Changes in protein constitution after infection with tobacco mosaic virus. *Virology* 40: 199-211.

Van Loon, L.C. and Van Strien, E.A. (1999). The families of pathogenesis–related proteins, their activities and comparative analysis of PR-1 type proteins. *Physiological and Molecular Plant Pathology* 55: 85-97.

Velazhahan, R., Cole, K.C., Anuratha, C.S., and Muthukrishnan, S. (1998). Induction of thaumatin-like proteins (TLPs) in *Rhizoctonia solani*-infected rice and characterization of two new cDNA clones. *Physiologia Plantarum* 102: 21-29.

Velazhahan, R., Datta, S.K., and Muthukrishnan, S. (1999). The PR-5 family: Thaumatin-like proteins. In Datta, S.K. and Muthukrishnan, S. (Eds.). *Pathogenesis-Related Proteins in Plants* (pp. 107-130). New York: CRC Press.

Velazhahan, R., Jayaraj, J., Jeoung, J.-M., Liang, G.H., and Muthukrishnan, S. (2002). Purification and characterization of an antifungal thaumatin-like protein from sorghum leaves. *Journal of Plant Diseases and Protection* 109: 452-461.

Velazhahan, R., Jayaraj, J., Liang, G.H., and Muthukrishnan, S. (2003). Partial purification and N-terminal amino acid sequencing of a β-1,3-glucanase from sorghum leaves. *Biologia Plantarum* 46: 29-33.

Velazhahan, R., Radhajeyalakshmi, R., Thangavelu, R., and Muthukrishnan, S. (2001). An antifungal protein purified from pearl millet seeds shows sequence homology to lipid transfer proteins. *Biologia Plantarum* 44: 417-421.

Vidhyasekaran, P. (1997). Fungal pathogenesis in plants and crops. In *Molecular Biology and Host Defense Mechanisms* (pp. 264-279). New York: Marcel Decker.

Vigers, A.J., Roberts, W.K., and Selitrennikoff, C.P. (1991). A new family of plant antifungal proteins. *Molecular Plant-Microbe Interactions* 4: 315-323.

Wooben, J.P., Lawrence, C.B., and De Wit, P.J.G.M. (1996). Differential induction of chitinase and β-1,3-glucanase gene expression in tomato by *Cladosporium fulvum* and its race-specific elicitors. *Physiological and Molecular Plant Pathology* 48: 105-116.

Xu, Y., Chang, Pi-F.L., Liu, D., Narasimhan, M. L., Raghothama, K.G., Hasegawa, P.M., and Bressan, R.A. (1994). Plant defense genes are synergistically induced by ethylene and methyl jasmonate. *Plant Cell* 6: 1077-1085.

Yamada, M. (1992). Lipid transfer proteins in plants and microorganisms. *Plant Cell Physiology* 33: 1-6.

Yedidia, I., Benhamou, N., Kapulnik, Y., and Chet, I. (2000). Induction and accumulation of PR proteins activity during early stages of root colonization by the mycoparasite *Trichoderma harzianum* strain T-203. *Plant Physiology and Biochemistry* 38: 863-873.

Yoshikawa, M., Tsuda, M, and Takeuchi, Y. (1993). Resistance to fungal diseases in transgenic tobacco plants expressing phytoalexin elicitor-releasing factor, β-1,3-endoglucanase from soybean. *Naturwissenschaften* 80: 417.

Zehnder, G.W., Murphy, J.F., Sikora, E.J., and Kloepper, J.W. (2001). Application of rhizobacteria for disease resistance. *European Journal of Plant Pathology* 107: 39-50.

Zhou, J.M. (1999). Signal transduction and pathogen-induced PR gene expression. In Datta, S.K. and Muthukrishnan, S. (Eds.). *Pathogenesis-Related Proteins in Plants* (pp. 195-206). New York: CRC Press.

Zhu, B., Chen, T.H.H., and Li, P.H. (1996). Analysis of late-blight disease resistance and freezing tolerance in transgenic potato plants expressing sense and antisense genes for an osmotin-like protein. *Planta* 198: 70-77.

Zhu, H., Krishnaveni, S., Liang, G.H., and Muthukrishnan, S. (1998). Biolistic transformation of sorghum using a rice chitinase gene. *Journal of Genetics and Breeding* 52: 243-252.

Zhu, Q., Maher, E.A., and Masoud, S. (1994). Enhanced protection against fungal attack by constitutive co-expression of chitinase and glucanase genes in transgenic tobacco. *Nature Biotechnology* 12: 807-812.

Ziadi, S., Barbedette, S., Godard, J.F., Monot, C., Le Corre, D., and Silue, D. (2001). Production of pathogenesis related proteins in the cauliflower (*Brassica oleracea* var. *botrytis*)-downy mildew *(Peronospora parasitica)* pathosystem treated with acibenzolar-S-methyl. *Plant Pathology* 50: 579-586.

Zimmerli, L., Jakab, G., Metraux, J.P., and Mauch-Mani, B. (2000). Potentiation of pathogen-specific defense mechanisms in *Arabidopsis* by beta-amino butyric acid. *Proceedings of the National Academy of Sciences, USA* 97: 12920-12925.

Chapter 6

Induced Plant Resistance to Fungal Pathogens: Mechanisms and Practical Applications

Ray Hammerschmidt

INTRODUCTION

It is commonly known that plants are resistant to the majority of pathogens found in the environment (Lucas, 1998). This suggests that all plants have the mechanisms necessary to effectively defend themselves against infection and that susceptibility may be the result of a failure of a plant to activate its defenses after infection or the ability of the pathogen to suppress or nullify the defenses. The fact that certain types of infections or chemical treatments can induce resistance in susceptible plants to pathogens is, therefore, not surprising because the capacity to resist infection is apparently universal in all plants.

In this chapter, the phenomenon of induced disease resistance to fungal pathogens will be discussed. Induced resistance is the phenomenon in which prior treatment changes the biochemistry or physiology of the plant in such a way that it is able to effectively resist infection by a wide range of pathogens (Hammerschmidt, 1999a). One of the hallmarks of induced resistance is the nonspecific nature of the resistance: once induced, the plant is often able to resist infection by multiple species of fungi, as well as other microbial pathogens. An-

The support of the Michigan Agricultural Experiment Station, USDA/CSREES, and the Michigan Soybean Promotion Committee for the author's research discussed in this chapter is gratefully acknowledged. I would also like to thank Rebecca Cifaldi for the preparation of Figure 6.1.

other characteristic is the lag time needed between the induction of treatment and expression of resistance. This time is needed to allow the development of the resistant state. The induced resistance is also often systemic in nature and can protect most, if not all, plant parts from infection. Figure 6.1 shows a generalized model of biologically induced resistance. In brief, treatment of a plant part (in this case a leaf) with a biological inducing agent results in the generation of a

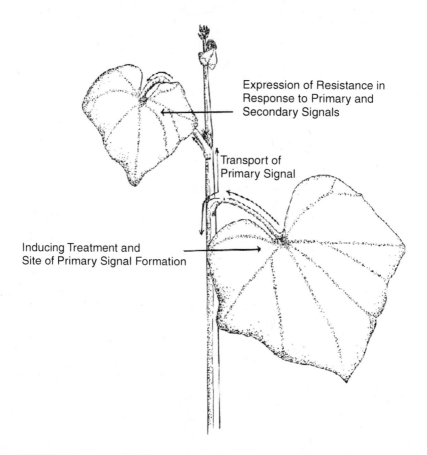

FIGURE 6.1. A general diagrammatic model of systemic induced/acquired resistance. An inducing pathogen is applied to the first true leaf, and its interaction with the plant generates a systemically translocated primary signal. This primary signal then induces a secondary signal that elicits the resistant state. Figure prepared by R. Cifaldi.

primary signal that is transported throughout the plant. This signal then has the capacity to induce other signals throughout the plant that are responsible for the activation of genes involved in the expression of induced resistance. This model also accommodates inducing agents that are abiotic but which also act through the induction of resistance-inducing signal models. Because of its broad spectrum and systemic nature, induced resistance has received attention as a potential disease-control measure. Induced-resistance model systems, such as cucumber, tobacco, and *Arabidopsis* (Hammerschmidt and Dann, 1997; Hammerschmidt, 1999a; Métraux et al., 2002; Sticher et al., 1997; Van Loon et al., 1998), have also provided systems to study the generation of systemic signals that control resistance, genes that regulate the expression of resistance, and genes that may be involved in plant defense. Thus, the phenomenon of induced resistance has become a valuable area of study and, as will be discussed later, a potential disease-management tool.

BACKGROUND

The phenomenon of induced resistance has been known since the early twentieth century. By 1933, enough evidence for the phenomenon had accumulated to allow the first comprehensive review of this topic by Chester (1933). This review was prophetic in that he suggested that induced resistance could be of practical importance and an area for fundamental studies on plant resistance.

Induced resistance can be either localized to the site of the inducing treatment or spread systemically throughout the plant. In 1940, Müller and Borger (reviewed in Hammerschmidt, 1999b) demonstrated that potato tuber tissue previously inoculated with an incompatible isolate of *Phytophthora infestans* would resist a subsequent infection, at the same site, by a virulent race of *P. infestans* or a tuber-infecting *Fusarium* species. Although these results were used by Müller and Borger in the development of the phytoalexin hypothesis, the study also clearly showed that resistance to pathogens can be induced in tissues that are susceptible to that pathogen. Twenty years later, Ross (1961) demonstrated that infection of N gene-containing tobacco with tobacco mosaic virus (TMV) would result in a systemic induction of resistance to subsequent inoculations with this virus. Later research showed that the systemic resistance induced by TMV was also effective against fungi and oomycetes (Hammerschmidt and Dann, 1997; Sticher et al., 1997).

In the 1970s, research by Kuc and colleagues began to build the framework for studies on induced resistance to disease and especially to fungal pathogens. They demonstrated that systemic resistance could be induced in cucumber plants by and against several different fungal pathogens (Kuc, 1982). His group repeatedly found that only those pathogens that caused a localized necrotic lesion would induce resistance and that the resistance was nonspecific (i.e., effective against several different pathogens). The successful cucumber research was built upon earlier work with green bean by his group (Kuc, 1982) and provided the groundwork for developing tobacco and *Arabidopsis* as model systems for induced resistance.

The type of induced resistance described has become commonly known as *systemic acquired resistance* (SAR). This type of induced resistance has specific characteristics, such as a need for a necrotic lesion produced by a pathogen as the inducing agent, systemic expression of pathogenisis-related (PR) protein genes, and involvement of salicylic acid as part of the signaling process (Hammerschmidt, 1999a). Over the past few years, it has been demonstrated that there are other forms of induced resistance. One of these is known as *induced systemic resistance* (ISR) (Van Loon et al., 1998; Pieterse et al., 2001). This type of induced resistance differs from SAR in that it is induced by certain rhizobacteria that do not cause a necrotic lesion and the resistance induction is not associated with PR protein accumulation. ISR also depends upon a perception of ethylene and jasmonic acid rather than salicylic acid as signals for resistance expression.

MECHANISMS OF INDUCED RESISTANCE TO FUNGAL PATHOGENS

Genetic Regulation

At least two forms of systemic induced resistance have been revealed by genetic and biochemical analysis: SAR and ISR. Much of this research has been carried out with *Arabidopsis* in which ease of genetic manipulation has allowed for rapid progress. Many excellent recent reviews provide an entry point into the rapidly growing research into the genetic regulation of induced resistance (e.g., Métraux et al., 2002) and thus this section will cover only the major points.

In SAR, an initial necrotic lesion results in the generation of a signal in the initially infected plant part. This signal is then translocated through the plant, may generate secondary signals, and results in the expression of pathogenesis-related protein genes and broad-spectrum resistance (Figure 6.1) (Sticher et al., 1997; Hammerschmidt, 1999a; Métraux et al., 2002).

One of the signals that the SAR response requires is salicylic acid (Figure 6.2; Métraux and Nawrath, 2002). Although some data suggest that salicylic acid is induced by another primary signal and is not the mobile signal, evidence that salicylic acid may be a primary signal also exists (see Métraux, 2002, for a brief review). However, experiments with tobacco and *Arabidopsis* transformed with *nahG,* a gene that codes for an enzyme that converts salicylic acid to catechol, have very clearly demonstrated the essential role of salicylic acid in SAR, even if its position in systemic signaling is still not clear (Gaffney et al., 1993; Delaney et al., 1994).

The use of *Arabidopsis* mutants has shed considerable light on the genetic regulation of SAR (Glazebrook, 2001), and a brief overview of the pathway from pathogen-induced necrosis to expression of resistance is shown in Figure 6.3. Genes such as *EDS1* (enhanced disease susceptibility) appear to be involved in the regulation of SAR, while further downstream, *EDS5/SID1* (salicylic acid induction deficient) and *SID2* appear to be more directly involved in salicylic acid biosynthesis or accumulation (reviewed in Métraux, 2002). *NPR1/NIM1* (no PR-1 expression; noninducible immunity) is required for regula-

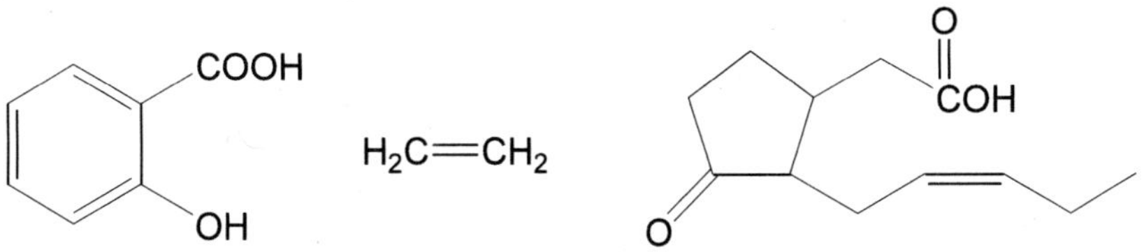

FIGURE 6.2. Three naturally occurring disease resistance signal molecules

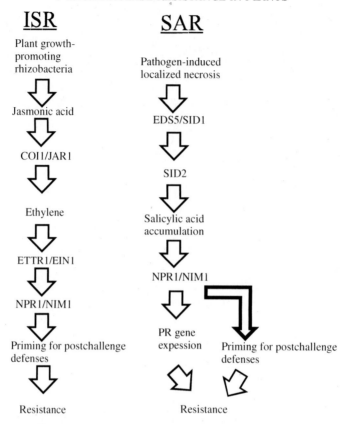

FIGURE 6.3. Signal pathways for ISR (induced systemic resistance) and SAR (systemic acquired resistance) (*Source:* Modified from Métraux, 2002, and Van Loon et al., 1998).

tion of SAR events downstream of salicylic acid and is needed for the expression of resistance and PR-protein gene expression (Cao et al., 1997; Ryals et al., 1997). The *NPR1/NIM1* protein is thought to activate expression of PR genes by interacting with basic leucine zipper transcription factors that, in turn, bind those sequences required for PR-1 gene induction by salicylic acid (Zhang et al., 1999).

Although salicylic acid is now well established as a key signal molecule in the regulation of SAR, the nature of its biosynthesis is not totally clear. For several years, evidence had been presented that salicylic acid was a product of the phenylpropanoid pathway and thus shared common phenolic compound intermediates such as cinnamic acid (Métraux, 2002; Figure 6.4). However, it has recently been shown that *SID2*, which was known to be involved in the accumulation of salicylic acid based on mutational and genetic mapping analyses, codes for a key enzyme in the biosynthesis of this phenolic acid. What was surprising about this result is that *SID2* appears to code for an isochorismate synthase (ICS) and not an enzyme in the phenylpropanoid pathway (Wildermuth et al., 2001). ICS converts the shikimic acid pathway intermediate chorismic acid into isochorismic acid which, in turn, can then be converted into salicylic acid (Figure 6.4). Interestingly, this is the route that prokaryotes use in the synthesis of salicylic acid, but the fact that many plants produce gallic acid (trihydroxybenzoic acid) directly from shikimic acid pathway inter-

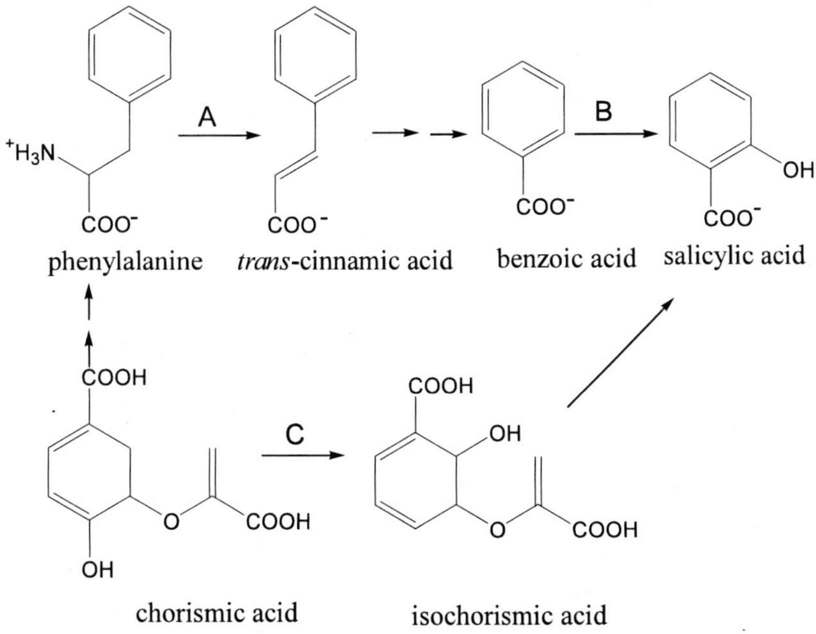

FIGURE 6.4. Biosynthesis of salicylic acid from a common shikimic acid pathway intermediate, chorismic acid: (A) phenylalanine ammonia lyase; (B) benzoic acid 4-hydroxylase; (C) isochorsimate synthase (*Source:* Modified from Métraux, 2002, and Wildermuth et al., 2001).

mediates does show that this type of pathway is not atypical for plants (Haslam, 1983).

ISR is induced by certain soilborne, plant-growth-promoting rhizosphere (PGPR) bacteria (Van Loon et al., 1998; Pieterse et al., 2001). Based on studies in *Arabidopsis,* this type of induced resistance differs from SAR in that no necrotic host response is needed to trigger the phenomenon; the pathway that leads to the expression of ISR is dependent on the ability to respond to ethylene and jasmonic acid (Pieterse et al., 1998, 2000, 2001) (Figure 6.2) (and not salicylic acid), and PR protein genes are not expressed (Pieterse et al., 1998; Van Loon et al., 1998). A proposed signaling pathway for ISR is shown in Figure 6.3. Interestingly, and perhaps not too surprisingly, ISR and SAR are effective against a slightly different range of pathogens (Ton et al., 2002). In these experiments, they found that SAR (induced by INA) in *Arabidopsis* (Columbia ecotype) was more effective against *Peronospora parasitica* than ISR induced by *P. fluorescens* WCS417r. These differences were observed for both the number of infected leaves and the amount of sporulation on the leaves. It is, however, of interest to note that none of the inducer treatments eliminated sporulation. On the contrary, *P. fluorescens* WCS417r induced resistance to the more necrotrophic fungus *Alternaria brassicicola,* in the *A. brassicicola*–susceptible pad-3 mutant of the Columbia ecotype of *Arabidopsis,* whereas inoculation with the SAR inducer *Pseudomonas syringae* pv. *tomato* did not protect the plants. In support of this differential response, methyl jasmonate, but not INA, protected the plants against *A. brassicicola.* Thus, it is possible that the two forms of induced resistance may not be compatible or have signaling pathways that interfere with each other. However, if SAR and ISR are induced simultaneously, the resistance is additive and there appears to be no negative influence of one type of resistance on the other (Van Wees et al., 2000; Pieterse et al., 2001).

Postinduction/Prechallenge Defenses

As a result of resistance induction, it is logical to assume that genes involved in the enhanced state of resistance have been induced or are expressed to a greater extent. In the case of the SAR form of induced resistance, this is clearly shown by the local and systemic expression of PR genes and accumulation of PR proteins (Van Loon, 1997; Hammerschmidt, 1999a). The very fact that many of these proteins have been shown to have antifungal activity or can degrade fungal cell walls in vitro suggests that the PR proteins may play a role in defense (Van Loon, 1997). In addition, the location of the acidic PR proteins in the apoplast indicates that the invading fungal hyphae will come into direct contact with these putative defense proteins (Van Loon, 1997). However, little direct evidence indicates that the presence of these proteins contributes directly to the induced resistance phenotype, although it is known that overexpression of certain PR-protein genes can enhance fungal disease resistance in some, but not all, cases (Punja, 2001).

Although it is now well known that the induction of SAR results in the systemic accumulation of salicylic acid (Métraux, 2002), little else has been reported on systemic changes in other secondary metabolites that may be involved in resistance. The potential role for defensive secondary metabolites accumulating systemically throughout plant tissues is known for resistance to insect herbivores (Hammerschmidt and Schultz, 1996), and thus it seems likely that this may also be true for systemic resistance to pathogens.

Recently, Prats and colleagues (2002) reported that prior treatment of sunflower plants with the resistance activator acibenzolar-*S*-methyl (ASM) induced resistance to *Puccinia helianthi* and also resulted in elevated levels of antifungal coumarins on the leaf surface. Cytological studies showed that induced plants supported lower uredospore germination and appressorium formation rates, suggesting that the coumarins were acting to reduce preinfection events. Examining other plants for acibenzolar-*S*-methyl–induced changes in secondary metabolites, as well as PR proteins, may be worth pursuing.

Other materials that have been shown to induce resistance may also act by increasing the content of antifungal compounds in the plant. For example, treatment of soybean plants with the diphenyl ether herbicide lactofen resulted in the accumulation of the phytoalexin glyceollin and increased resistance to *Sclerotinia sclerotiorum* (Dann et al., 1999). The herbicide may be triggering the resistance response through the accumulation of protoporphyrins which generate active oxygen species in the light (Hammerschmidt and Dann, 1999).

Treatment of cucumber plants with soluble silicon or an extract of the giant knotweed *(Reynoutria sachalinensis),* or with plant-growth-promoting rhizobacteria, induced resistance to powdery mildew on foliage or *Pythium* on roots, respectively (Daayf et al., 1997; Fawe et al., 1998; Ongena et al., 2000). In all three cases, the inducing treatments resulted in the accumulation of glycosides of antifungal phenolic compounds (both simple hydroxycinnamic acids and a flavonol). Although the aglycones released by acid hydrolysis of the glycosides were more antifungal than the parent compounds, it is possible that the enhanced glycosides served as a stored source of defensive compounds that were released by glycosidases after the host was infected, and this would be similar to the release of cyanide or isothiocyanates from cyanogenic glycosides or glucosinolates, respectively, after damage to plant cells caused by insects (Renwick, 1997). Enhanced hydrolases (glucanase and chitinase) have been reported in pea plants in relation to silicon-induced resistance to fungi (Dann and Muir, 2002), and it would be of interest to see if the aforementioned inducing treatments also induced an increase in glycosidases capable of releasing toxic phenolic compounds.

Postchallenge Induction of Defenses

In 1982, Hammerschmidt and Kuc reported that cucumber plants expressing systemic induced resistance to *Colletotrichum orbiculare* and *Cladosporium cucumerinum* deposited lignin more readily at the point of fungal infection into induced plants when compared to the control. Other examples of defenses (e.g., phytoalexins, cell wall alterations, hypersensitivity) have also been observed in induced plants following inoculation with fungal pathogens (Hammerschmidt, 1999b). The examples cited are SAR-type induced resistance, but an increased capacity to synthesize phytoalexins in plants expressing ISR is also known. Van Peer and colleagues (1991) demonstrated that *Pseudomonas* strain WCS417r, a rhizobacterium, protected carnation against *Fusarium,* and this protection was associated with accumulation of phytoalexins after attempted infection by the pathogen. These reports suggested that the induction treatment had, in some way, prepared noninfected tissue to respond more quickly to subsequent challenge infections.

The enhanced ability to express defenses after challenge is now generally known as *priming,* and other good examples of this are available in the literature (Conrath et al., 2002). Although the molecular basis for priming has not been established, a requirement for *NPR1/NIM1* has been shown (Kohler et al., 2002). Treatment of *Arabidopsis npr1/nim1* mutants with acibenzolar-*S*-methyl (ASM) did not condition the plants to rapidly express the phenylalanine ammonia lyase (PAL) gene after inoculation or to more rapidly deposit callose after gentle wounding. Thus, understanding how (or if) the NPR1/NIM1 protein interacts with defense processes other than PR gene expression may be key to determining how the primed state becomes established.

A report by Latunde-Dada and Lucas (2001) further supports a role for priming in the expression of defenses after challenge inoculations. ASM-induced resistance in cowpea to *Colletotrichum destructivum* was shown to be associated with several postchallenge biochemical changes. Inoculation of induced plants with *C. destructivum* resulted in the expression of a hypersensitive-like response in the invaded host cell and failure of the pathogen to develop out of the initially infected cell. The accumulation of the phytoalexin phaseollidin occurred much more rapidly and to a greater concentration after fungal challenge in induced tissues compared to the controls. The activity of phenylalanine ammonia lyase and chalcone isomerase, two enzymes involved in synthesis of the phytoalexin, also increased after infection and, as would be expected, prior to the accumulation of phaseollidin.

Taking a different approach, Stadnik and Buchenauer (2000) showed that inhibiting the enzyme phenylalanine ammonia lyase suppressed ASM-induced resistance to *Blumeria graminis* f. sp. *tritici* in wheat. Treating induced leaf tissues with the PAL inhibitor α-amino-β-phenylpropionic acid (AOPP) reduced autofluorescence associated with failed penetration sites and decreased the level of induced resistance. AOPP treatment also reduced accumulation of wall-esterified *p*-coumaric and ferulic acids. Treatment with an inhibitor of cinnamyl alcohol dehydrogenase, an enzyme involved in lignin biosynthesis, did not reduce the level of induced resistance.

We have recently observed that cycloheximide suppresses induced resistance in cucumber leaves to *C. orbiculare* (Velasquez and Hammerschmidt, unpublished data). The cycloheximide-treated induced plants allowed greater penetration of the pathogen into the tissue than did induced plants not treated with the protein synthesis inhibitor, suggesting that new protein synthesis was needed after challenge to block fungal penetration. The activity of chitinase in the induced leaf tissues remained high after challenge, even in the presence of cycloheximide. Interestingly, the rate of hyphal growth in the induced tissue in the presence or absence of cycloheximide was similar, suggesting that chitinase alone was not sufficient to stop fungal development. This further indicates a need for additional defenses to be activated after challenge.

Why Are Induced Plants More Resistant to Fungi?

The previous two sections briefly described some of the putative plant defense responses associated with induced resistance to fungi. However, two questions are still unanswered: How do the induced plants stop development of the fungi and, in the case where post-challenge defenses are induced, how does the induced plant recognize the presence of the pathogen? The first question was addressed in a recent review (Hammerschmidt, 1999b), and the second question remains largely unanswered. However, research tools in the form of antisense expression or sense suppression of key defense genes, or the use of fungi that are tolerant of defenses such as phytoalexins, may begin to provide these answers for induced resistance and other forms of active defenses as has been used in the study of phytoalexins (Hammerschmidt, 1999b). Certainly, the lessons learned from examining plants that are overexpressing PR genes (Punja, 2001) suggest that these proteins are only part of the defenses that account for the induced resistance phenotype. The second question, however, may be intriguing because the formerly susceptible plant appears to now recognize the fungal pathogen and mounts a rapid response. Where examined, induced plant cells respond actively to attempted infection by fungi, but it is not at all known why or how this occurs or how the plant recognizes the pathogen. The observation that induced plant cells are more responsive to fungal elicitors may provide one explanation (Katz et al., 1998; Krauss et al., 1993).

Because induced resistance to fungi is not specific to any one species, the detection of these fungal invaders by enhanced responsiveness to elicitors may help account for the rapid deployment of additional defenses. In the case of SAR, the induction phase results in the systemic expression of PR proteins, such as chitinases and glucanases, that accumulate in the apoplast (Van Loon, 1997). Because the enzymes are in the apoplast, they should come into contact with the invading hyphae and release cell wall oligomers that can act as elicitors (Lawrence et al., 2000) which could then trigger an active defense response in the already primed tissues. Receptors for fungal cell wall elicitors have been identified (Cote et al., 2000), and the relation of these to resistance expression could be worth exploring.

The release of elicitors from hyphae by hydrolytic PR proteins is one possible way that a plant may recognize the presence of a pathogen. However, since not all forms of induced resistance are associated with PR proteins, this cannot be the only mechanism of recognition. Treatment of rice with resistance-inducing compounds has been shown to induce the expression of a gene, *RPR1* (Sakamoto et al., 1999). This gene is similar to several known R genes in that the protein it codes for has a nucleotide binding site and leucine-rich repeats. Although it is only pure speculation, this result suggests that perhaps genes that have similar properties to R genes may be involved in recognition of pathogens by induced plants and thus allow for the expression of genes involved in defense.

THE APPLICATION OF INDUCED RESISTANCE

The broad-spectrum nature of induced resistance to fungal pathogens makes this type of resistance a potentially useful approach to disease management. Early field research by Kuc and colleagues (Kuc, 1982) demonstrated that field-grown cucumber plants could be protected against infection by *C. orbiculare* by a prior localized infection by the same pathogen. Although the approach worked, the need to individually inoculate plants makes any practical application of this method highly impractical for modern production agriculture. For induced resistance to have practical application, the delivery of inducing agents must be compatible with current agricultural practices, be easy to apply, and be reliable. In this section, a few selected resistance-inducing agents and their uses for fungal disease control will be discussed.

The practical application of induced resistance to crop protection requires that some type of product (biotic or abiotic) be used that can be easily applied to the plant and is also easily integrated into standard disease management protocols. In 1994, Kessmann and colleagues developed a set of criteria to determine whether a material is an inducer or activator of resistance. Although the characterization of ISR and SAR has shown that there are likely multiple forms of induced resistance, the strict criteria of Kessmann and colleagues (1994) may not be universally applicable (they based their criteria on the SAR type of induced resistance). It is important that any product or material being developed as a resistance activator must act via enhancing plant resistance and not as an antimicrobial compound. In this section, a few types of materials that are known to induce disease resistance will be discussed.

Salicylic Acid and Its Functional Analogs

Ever since White (1979) found that treatment of tobacco plants with salicylic acid or aspirin would induce disease resistance, it was clear that the external application of a chemical could be used to protect plants by enhancing disease resistance. Based on the principles of induced disease resistance and what properties a resistance activator should have, the first synthetic resistance inducers were developed (Tally et al., 1999). The first of these, 2,6-dichloroisonicotinic acid (Figure 6.5), was capable of inducing resistance in a number of crops against a fairly wide range of pathogens and had characteristics that suggested it functioned as an analog of salicylic acid (Kessmann et al., 1994). Another functional analog of salicylic acid, ASM (Figure 6.5), has been developed into a commercial product (Actigard or Boost) that is registered on several crops (Tally et al., 1999). ASM has been shown to be effective as a foliar spray (Oostendorp et al., 2001; Tally et al., 1999), as well as a seed treatment (Jensen et al., 1998). Both INA and ASM reduced the severity of *Sclerotinia* stem rot on soybean under field conditions (Dann et al., 1998). However, even though the severity of disease was often significantly reduced and the reduction was seen in several cultivars that differed in resistance to this pathogen, the need for multiple applications and a failure to realize increased yields along with disease suppression suggest that this type of activator may not be suitable for all plant-patho-

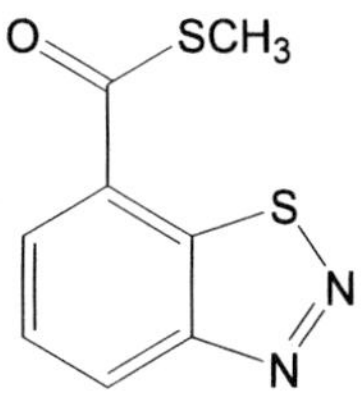

Acibenzolar-S-methyl
(ASM, Actigard, Boost)

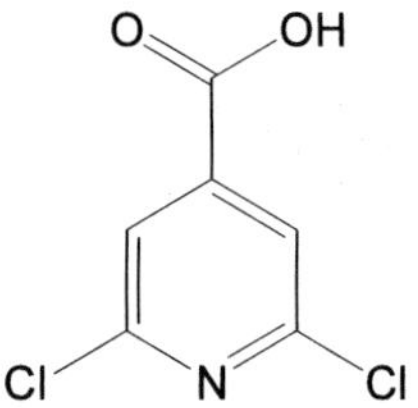

2,6-dichloroisonicotinic acid
(INA)

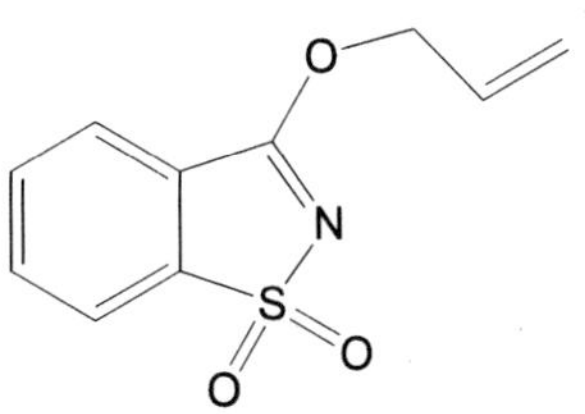

3-allyloxy-1,2-benzisothiazole-1,1-dioxide
(Probenazole)

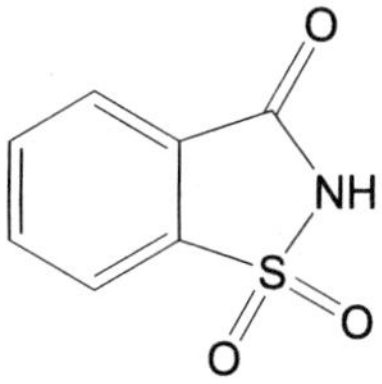

1,2-benzisothiazol-3(2H)-one-1,1-dioxide
(BIT, saccharin)

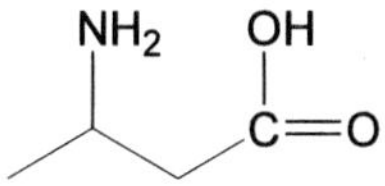

DL-3-amino-n-butyric acid
(BABA)

FIGURE 6.5. Synthetic disease resistance inducing/activating compounds

systems. Because there is a lag time between when ASM is applied to plants and when resistance is expressed (something common to all forms of induced resistance), the application of ASM must be made prior to an infection event and/or applied in a formulation with a protectant fungicide (Oostendorp et al., 2001).

Nonprotein Amino Acids

Over the past ten years, research by Cohen and others has shown β-amino-butyric acid (BABA) (Figure 6.5) to be an effective inducer of disease resistance against fungal pathogens in several crop species and it is being intensively investigated (see review by Cohen, 2002). Although some debate surrounds how BABA can induce resistance (Cohen, 2002), it is clear from Cohen's exhaustive review of the literature that this material does induce PR protein accumulation in some plant species and conditions many plant species to respond more quickly to attempted infection. For example, pepper stems pretreated with BABA accumulated chitinase and glucanse, and the treated plant tissues were primed to more rapidly produce phytoalexins after inoculation with *Phytophthora capsici* as compared to controls (Hwang et al., 1997). Both field and greenhouse trials have shown that BABA has potential as a crop protection material that can be applied as a foliar spray, seed treatment, or soil drench and can also be applied via drip irrigation (see Cohen, 2002).

Organic and Inorganic Naturally Occurring Compounds

Oxalic acid and sodium phosphates have been shown to reduce disease severity (Oostendorp et al., 2001; Pajot et al., 2001). Silicon oxides have also been reported to enhance disease resistance in cucumber to powdery mildew (Fawe et al., 1998) and in pea plants to *Mycosphaerella pinodes* (Dann and Muir, 2002).

Fungicides

Some reports in the literature suggest that the mode of action of certain fungicides is through the induction of resistance in the plant (Oostendorp et al., 2001). However, only limited information that suggests fungicides directly act as activators of resistance. The fungicide probenazole (Figure 6.5) has true fungicidal properties (cited in Oostendorp et al., 2001) but has also been recently found to act directly and through a metabolite [1,2-benzisothiazol-3 (2H)-one-1,1-dioxide or BIT, Figure 6.5] to be an inducer of resistance (Yoshioka et al., 2001). Treatment of *Arabidopsis* with probenazole or BIT induced resistance to *Peronospora parasitica* as well as induced the accumulation of salicylic acid and PR proteins. The induced resistance appeared to follow a SAR type of pathway, and both salicylic acid and *NPR1* were shown to be required.

Certain fungicides, however, may indirectly activate certain aspects of host plant defenses. Ward and colleagues published several papers indicating that prior treatment of soybean seedlings with metalaxyl resulted in the expression of a resistance-like response after inoculation with a compatible isolate of *Phytophthora sojae* (Ward, 1984). Associated with this response was the accumulation of the phytoalexin glyceollin (Ward, 1984). Although the accumulation of glyceollin was not sufficient to account for the failure of the pathogen to colonize the tissue, the necrotic lesion response and phytoalexin induction indicated that a localized defense response was induced. The major role that metalaxyl played in protecting the plants from virulent isolates was demonstrated by the failure of metalaxyl-treated plants to resist or accumulate glyceollin when inoculated with metalaxyl-resistant isolates of the pathogen.

The failure of the metalaxyl-resistant isolates to elicit a defense response in metalaxyl-treated tissue may be due to the lack of elicitor release by these isolates. Treatment of metalaxyl-sensitive isolates of *Phytophthora* with metalaxyl resulted in the release of glyceollin elicitors, while resistant strains did not release elicitors (Cahill and Ward, 1989). Thus, it is possible that when metalaxyl-treated soybean seedlings were infected with virulent, metalaxyl-sensitive isolates of the pathogen, the metalaxyl had direct antifungal activity on the pathogen that stopped the infection, but the stressed or dying pathogen then released elicitors that induced the localized resistance response.

Protection of *Arabidopsis* plants against *Peronospora parastica* with several different fungicides (CuOH, metalaxyl, and fosetyl-Al) was not as effective in plants expressing *nahG* as compared to wild type (Molina et al., 1998). The authors suggested that a normal host defense signaling pathway must be functioning to allow full action of these fungicides. Although the manner in which the fungicides interacted in this case was not elucidated, the data are in line with the conclusions of Ward and colleagues that certain fungicides may act in concert with host defenses via an indirect mode of action.

Taken together, these results suggest that most fungicides do not directly induce resistance, but that their action on the pathogen may enhance localized resistance via release of elicitors. To fully test the role of fungicides as inducers, it is important to test for plant resistance with isolates of the pathogen that are resistant to the fungicide and to any antifungal metabolites of the compound that may be present in the plant to determine if the plants are less diseased when infected with a resistant isolate in combination with the fungicide as compared to a non-fungicide-treated control. The value of using resistant isolates in examining the role of fungicides as inducers of resistance was clearly illustrated in the soybean research in which isolates of *Phytophthora* that were metalaxyl-resistant parasitized tissues that were treated or untreated with metalaxyl with comparable virulence. Because the presence of the pathogen in the fungicide-treated plants does elicit localized defense responses and necrosis, it is possible that the fungicide-mediated host response may trigger a SAR response.

Herbicides

Another group of common agrochemicals, the herbicides, have also been reported to enhance or induce disease resistance. Dann and colleagues (1999) found that field or greenhouse treatment of soybean with the diphenylether herbicide lactofen reduced the severity of *Sclerotinia* stem rot and, when disease pressures were high, sometimes resulted in an increase in yield. The herbicidal effect of lactofen is based on its ability to block protoporphynogen oxidase, and this results in the accumulation of protoporphoryn which, in the presence of light, produces activated oxygen species (Hammerschmidt and Dann, 1997). This ability to produce activated oxygen may be related to the observation that lactofen induced the accumulation of the phytoalexin glyceollin. Thus, some of the protective effects of lactofen may be due to its abilility to induce phytoalexin synthesis.

Biological Inducers

Certain plant-growth-promoting rhizobacteria have been shown to be effective inducers of resistance in plants such as cucumber, tobacco, and *Arabidopsis* (Van Loon et al., 1998; Zhender et al., 2001). Where studied in detail, the type of resistance induced by PGPR has often been shown to be of the ISR type (Van Loon et al., 1998). Because the PGPR can be applied directly to the soil (Van Loon et al, 1998) or used as a seed treatment (Zhender et al., 2001), the use of PGPR as crop protection tools seems very possible.

Combining Resistance Inducers with Traditional Chemical Controls

A clear message that comes from reviewing over 100 papers on induced disease resistance is that in only a few cases (if any) can induced resistance provide complete control of a pathogen. Although this statement can be made for many other control measures, the fact remains that induced resistance will likely not be a stand-alone disease-management tool. For example, in the cucumber induced-resistance systems, the resistance to *Colletotrichum orbiculare* was based on a reduction of successful penetrations by the pathogen (Hammerschmidt and Kuc, 1982). Induced resistance also requires time to develop, and it generally cannot cure established infections. Thus, the plant can be vulnerable to infection between the time that the plant activator or inducer is applied and when induced resistance becomes established. Finally, we have found that, in cucumber, not all cultivars respond equally to induction by the activator ASM (Velasquez, DaRocha, and Hammerschmidt, unpublished), and thus the plant genotype itself may dictate how effective induced resistance will be. One solution to this is to combine the use of plant resistance activators with standard fungicides to protect the plant during the induction phase. Oostendorp and colleagues (2001) presented data which showed that disease control exhibited by fungicides can be enhanced by activators such as ASM, and vice versa. Cohen (2002) reported that BABA could enhance the efficacy of fungicides even when BABA alone was inactive. When considered in light of the previous discussion on fungicides as direct or indirect inducers of resistance, these results are not surprising and indicate that more research to evaluate the combined effects of plant resistance activators or inducers and fungicides has merit.

Pathogen resistance is a problem that faces the future use of many fungicides, and it is possible that the use of plant activators may help reduce this risk. Treatment of apple seedlings with INA resulted in induced resistance to the apple scab fungus, *Venturia inaequalis* (Ortega et al., 1998). A dose-response study of the fungicide flusilazole in induced apple seedlings revealed a synergistic effect between the fungicide and INA. The authors hypothesized that using plant activators could reduce the number of applications or concentrations of fungicides used and thus reduce the risk of developing fungicide resistance. A synergistic effect between ASM and three different fungicides (metalaxyl, fosetyl-Al, and CuOH) was also demonstrated in the protection of *Arabidopsis* against *P. parasitica* (Friedrich et al., 2001). Thus, it may not be unreasonable to hypothesize that in induced plants treated with fungicides, more elicitors may be released from fungicide-weakened fungi by chitinases and glucanases, and this may act to further stimulate induced resistance defenses that are already primed for expression.

TRANSGENIC PLANTS
AND INDUCED RESISTANCE TO FUNGI

Plant transformation methods have provided tools to study the nature of induced resistance as well as to provide potential applications of knowledge learned through basic studies of induced resistance. The use of PR genes in the transformation of plants to enhance resistance to fungi was one of the first uses of transgenic plants in the study and application of induced resistance. For example, transformation of tobacco with the PR1-gene (Alexander et al., 1993) resulted in plants that had decreased susceptibility to *Peronospora tabacina*. In most cases where PR genes or other genes associated with induced resistance (PR genes, peroxidases, etc.) have been used to generate transgenic plants, the effect on fungal pathogens has been similar to that just described: either a small reduction in disease or no effect (Punja, 2001). However, because in most cases we do not know the *in planta* role of these genes and because plant resistance is based on multiple defenses, these reports may indicate the level of participation that these genes have in defense.

As previously discussed, transformation of tobacco and *Arabidopsis* with *nahG*, a bacterial gene that codes for a salicylate hydroxylase, was instrumental in determining the important role of salicylic acid in SAR by converting salicylate to an inactive metabolite (Gaffney et al., 1993; Delaney et al., 1994). Thus, if eliminating salicylic acid by transformation can eliminate SAR, it seems logical that resistance would be increased in plants with increasing amounts of salicylic acid. Mauch and colleagues (2001) transformed *Arabidopsis* with an isochorismate synthase gene and a pyruvate lyase gene (genes that convert chorismic acid to salicylic acid). Interestingly, a study by Wildermuth and colleagues (2001) showed that a major route for salicylic acid was via an isochorismate synthase. When the genes were targeted to the chloroplast, the plants were greatly stunted. However, when the genes were targeted to the cytosol, the plants appeared normal, salicylate content was elevated, and resistance to *Peronospora parasitica* was enhanced.

Mutations in the *NPR1/NIM1* gene resulted in plants that could not express either ISR or SAR (e.g., Delaney et al., 1995). However, overexpression of *NPR1* in *Arabidopsis* resulted in plants that expressed resistance to several pathogens (Cao et al., 1998), further demonstrating the important role of the protein product of this gene in resistance. Overexpression of *NIM1* in *Arabidopsis* also enhanced general disease resistance (Friedrich et al., 2001). Interestingly, the *NIM1* overexpressing plants exhibited a more rapid expression of *PR1* after infection with *P. parasitica* and were more responsive to treatment with the plant activator ASM. The overall control of *P. parasitica* by metalaxyl, fosetyl-Al, and CuOH was also enhanced in the NIM1 transgenics as compared to the wild-type plant, again supporting a possible synergistic activity between induced resistance and traditional fungicides. The authors suggested that the use of this type of transgenic could also allow for use of lower amounts of ASM to achieve full expression of induced resistance (Friedrich et al., 2001). It seems likely that combining the use of *NIM1* transgenics (or similar genes) in disease-management plans could reduce the input of both traditional chemical controls and plant activators.

CURRENT RESEARCH PROGRAM ON INDUCED RESISTANCE IN CUCUMBER

Current research in my laboratory is focusing on the genotypic variability of induced resistance in cucumber as well as differential forms of induced resistance against a hemibiotrophic fungus and a necrotroph. Treatment of a range of cucumber cultivars with ASM has induced varying degrees of resistance to infection by the hemi-biotrophic fungal pathogen *Colletotrichum orbiculare,* ranging from little induction to very high resistance. When the same cultivars were tested for induced resistance to *Didymella bryoniae,* some cultivars became more resistant, but several showed little response or became more susceptible. Analysis of chitinase induction by ASM treatment also showed a wide range of responses across the cultivars, but no correlation between the level of induced resistance and the induced activity of chitinase was found. Two cultivars, SMR 58 and Marketmore 86, were studied in more detail. Treatment with ASM, INA, and inoculation with *C. orbiculare* induced resistance to both pathogens. When 'Marketmore 86' was tested, all inducing agents induced resistance to *C. orbiculare,* but only *C. orbiculare* induced resistance to - *D. bryoniae.* Interestingly, both wounding and jasmonic acid treatment induced resistance to *D. bryoniae* but not to *C. orbiculare.* These results suggest that different signaling pathways also exist in cucumber plants, and the differential expression of resistance we have observed is similar to that previously discussed for *Arabidopsis* (Ton et al., 2002).

We have also examined a wide range of chemical and biologically based materials to test their efficacy as inducers of resistance in wounded potato tubers to *Fusarium sambucinum* (Greyerbiehl and Hammerschmidt, unpublished). Most compounds tested (ASM, BABA, salicylic acid, arachidonic acid) had little or no effect (even when tested in combinations). However, chitosan was quite effective as an inducer. Chitosan treatment induced the formation of what appeared to be a phenolic barrier near the surface of wounded potato tuber tissue, while at the same time suppressing the accumulation of wound-induced steroid glycoalkaloids. The response to chitosan was also systemic (over a few centimeters) into the tuber tissue with respect to chitinase induction.

CONCLUSIONS

Research on the phenomenon of induced disease resistance has literally exploded over the past decade. Tremendous advances have been made in our understanding of the genetic regulation of both the SAR and ISR forms of induced resistance, and this information has been useful in understanding resistance in general. Most of the molecular and genetic research has relied on the *Arabidopsis* system, and much of the research reported in this chapter comes from studies with *Arabidopsis.* However, it is increasingly more important that the knowledge learned about the regulation of induced resistance be used to study the regulation of induced resistance in other crops and to determine if what is observed in *Arabidopsis* can be applied directly or indirectly to other systems. Research on the regulation of SAR and

ISR has the potential for revealing new means of disease control in other crops, and this new knowledge should be used. The fact that induced resistance is manifested in more than one form further indicates that plants have sophisticated means of dealing with different types of plant pathogens, and understanding how the signaling pathways interact in both a positive and negative way is important if implementation of induced resistance as a disease-management tool is to become a reality.

A plethora of papers have described a wide variety of natural and synthetic compounds as well as biologicals that claim to induce resistance. In some well-studied cases (e.g., PGPR-mediated ISR, ASM, INA, BABA), very solid evidence indicates that these compounds are inducers of resistance. Other materials deserve equal study so that it can be determined if these are true inducers or only clandestine fungicides. Such studies on mode of action may, as was the case with the PGPRs, lead to new forms of induced resistance that are controlled by different signaling pathways. More effort on the field testing of the efficacy of inducers as stand-alone products and in combination with traditional chemicals, host resistance, and cultural practices are also needed to help find the proper fit for induced resistance, in disease management. Care should be taken to observe any obvious cultivar differences that may lead to differential expression of resistance within a species, as well as more studies that evaluate the effects of inducers on the yield and quality of the harvested product since just controlling the disease may not be enough.

REFERENCES

Alexander, D., Goodman R.M., Gut-Rella, M., Glascock, C., Weyman, K., Friedrich, L. Maddox, D., Ahl-Goy, P., Luntz, T., Ward, E., and Ryals, J. (1993). Increased tolerance to two oomycete pathogens in transgenic tobacco expressing pathogenesis-related protein 1a. *Proceedings of the National Academy of Sciences, USA* 90: 7327-7331.

Cahill, D.M. and Ward, E.W.B. (1989). Effects of metalaxyl on elicitor activity, stimulation of glyceollin production and growth of sensitive and tolerant isolates of *Phytophthora megasperma* f. sp. *glycinea*. *Physiological and Molecular Plant Pathology* 35: 97-112.

Cao, H., Glazebrook, J., Clarke, J.D., Volko, S., and Dong, X. (1997). The *Arabidopsis* NPR1 gene that controls systemic acquired resistance encodes a novel protein containing ankyrin repeats. *Cell* 88: 57-63.

Cao, H., Li, L., and Dong, X. (1998). Generation of broad spectrum resistance by overexpression of an essential regulatory gene in systemic acquired resistance. *Proceedings of the National Academy of Sciences, USA* 95: 6531-6536.

Chester, K.S. (1933). The problem of acquired physiological immunity in plants. *Quarterly Review of Biology* 8: 275-324.

Cohen, Y. (2002). β-Aminobutyric acid-induced resistance against pathogens. *Plant Disease* 86: 448-457.

Conrath, U., Pirtesse, C.M.J., and Mauch-Mani, B. (2002). Priming in plant-pathogen interactions. *Trends in Plant Science* 7: 210-216.

Cote, F., Roberts, K.A., and Hahn, M.G. (2000). Identification of high-affinity binding sites for the hepta-beta-glucoside elicitor in membranes of the model legumes *Medicago truncatula* and *Lotus japonicus. Planta* 211: 596-605.

Daayf, F., Schitt, A., and Belanger, R.R. (1997). Evidence of phytoalexins in cucumber leaves infected with powdery mildew following treatment with leaf extracts of *Reynoutria sachalinensi. Plant Physiology* 113: 719-727.

Dann, E.K., Diers, B., Byrum, J., and Hammerschmidt, R (1998). Effects of treating soybean with 2,6-dichloroisonicotinic acid and benzothiadiazole (BTH) on seed yields and level of disease caused *by Sclerotinia sclerotiorum* in field and greenhouse studies. *European Journal of Plant Pathology* 104: 271-278.

Dann, E.K., Diers, B.W., and Hammerschmidt, R. (1999). Suppression of *Sclerotinia* stem rot of soybean by lactofen herbicide treatment. *Phytopathology* 89: 598-602.

Dann, E.K. and Muir, S. (2002). Peas grown in media with elevated plant available silicon levels have higher activities of chitinase and β-1,3-glucanase are less susceptible to a fungal leaf pathogen, and accumulate more foliar silicon. *Australian Plant Pathology* 31: 9-13.

Delaney, T.P., Friedrich, L., and Ryals, J. (1995). *Arabidopsis* signal transduction mutant defective in chemically and biologically induced disease resistance. *Proceedings of the National Academy of Sciences, USA* 92: 6602-6606.

Delaney, T.P., Uknes, S., Vernooij, B., Friedrich, L., Weymann, K., Negrotto, D., Gaffney, T., Gut-Rella, M., Kessmann, F., Ward, E., and Ryals, J. (1994). A central role of salicylic acid in plant disease resistance. *Science* 266: 1247-1250.

Fawe, A., Abou-Zaid, M., Menzies, J.M., and Belanger, R.R. (1998). Silicon-mediated accumulation of flavonoid phytoalexins in cucumber. *Phytopathology* 88: 396-401.

Friedrich, L., Lawton, K., Dietrich, R., Willits, M., Cade R., and Ryals, J. (2001). *NIM1* overexpression in *Arabidopsis* potentiates plant disease resistance and results in enhanced effectiveness of fungicides. *Molecular Plant-Microbe Interactions* 14: 1114-1124.

Gaffney, T., Friedrich, L., Vernooij, B., Negrotto, D., Nye, G., Uknes, S., Ward, E., Kessmann, H., and Ryals, J. (1993). Requirement of salicylic acid for the induction of systemic acquired resistance. *Science* 261: 754-766.

Glazebrook, J. (2001). Genes controlling expression of defense responses in *Arabidopsis*—2001 status. *Current Opinion in Plant Biology* 4: 301-308.

Hammerschmidt, R. (1999a). Induced disease resistance: How do induced plants stop pathogens? *Physiological and Molecular Plant Pathology* 55: 77-84.

Hammerschmidt, R. (1999b). Phytoalexins: What have we learned after 60 years? *Annual Review of Phytopathology* 37: 285-306.

Hammerschmidt, R. and Dann, E.K. (1997). Induced resistance to disease. In Rechcigl, N.A. and Rechcigl, J.E. (Eds.), *Environmentally Safe Approaches to Plant Disease Control* (pp. 177-199). Boca Raton, FL: CRC Lewis Publishers.

Hammerschmidt R. and Dann E.K. (1999). The role of phytoalexins in plant protection. In Chadwick, D., and Goode, J.A. (Eds), *Insect-Plant Interactions and Induced Plant Defence* (pp. 175-190). Chichester, UK: Wiley.

Hammerschmidt, R. and Kuc, J. (1982). Lignification as a mechanism for induced systemic resistance in cucumber. *Physiological Plant Pathology* 20: 61-71.

Hammerschmidt, R. and Schultz, J. (1996). Multiple defenses and signals in plant defense against pathogens and herbivores. *Recent Advances in Phytochemistry* 30: 121-154.

Haslam, E. (1993). *Shikimic acid: Metabolism and metabolites*. New York: John Wiley and Sons.

Hwang, B.K., Sunwoo, J.Y., Kim, Y.J., and Kim, B.S. (1997). Accumulation of β-1,3-glucanase and chitinase isoforms and salicylic acid in the DL-β-amino-n-butyric acid-induced resistance response of pepper stems to *Phytophthora capsici*. *Physiological and Molecular Plant Pathology* 51: 305-322.

Jensen, B.D., Latunde-Dada, A.O., Hudson, D., and Lucas, J.A (1998). Protection of *Brassica* seedlings against downy mildew and damping-off by seed treatment with CGA 245704, an activator of systemic acquired resistance. *Pesticide Science* 52: 63-69.

Katz, V.A., Thulke, O.U., and Conrath, U. (1998). A benzothiadiazole primes parsley cells for augmented elicitation of defense responses. *Plant Physiology* 117: 1333-1339.

Kauss, H., Franke, R., Krause, K., Conrath, U., Jeblick, W., Grimming, B., and Matern, U. (1993). Conditioning of parsley *Petroselinum crispum* L. suspension cells increases elicitor-induced incorporation of cell wall phenolics. *Plant Physiology* 102: 459-466.

Kessmann, H., Staub, T., Hofmann, C., Maetzke, T., Herzog, J., Ward, E., Uknes, S., and Ryals, J. (1994). Induction of systemic acquired disease resistance in plants by chemicals. *Annual Review of Phytopathology* 32: 439-459.

Kohler, A., Schwindling, S., and Conrath, U. (2002). Benzothiadiazole-induced priming for potentiated responses to pathogen infection, wounding, and infiltration of water into leaves requires the *NPR1/NIM1* gene in *Arabidopsis*. *Plant Physiology* 128: 1046-1056.

Kuc, J. (1982). Induced immunity to plant disease. *BioScience* 32: 854-860.

Latunde-Dada, A.O. and Lucas, J.A. (2001). The plant defence activator acibenzolar-S-methyl primes cowpea [*Vigna unguiculata* (L.) Walp.] seedlings for rapid induction of resistance. *Physiological and Molecular Plant Pathology* 58: 19-208.

Lawrence, C.B. Singh, N.P., Qiu, J., Gardner, R.G., and Tuzun, S. (2000). Constitutive hydrolytic enzymes are associated with polygenic resistance of tomato to *Alternaria solani* and may function as an elicitor release mechanism. *Physiological and Molecular Plant Pathology* 7: 211-220.

Lucas, J.A. (1998). *Plant Pathology and Plant Pathogens.* Oxford: Blackwell Science.

Mauch, F., Mauch-Mani, B., Gaille, C., Kull, B., Haas, D., and Reimmann, C. (2001). Manipulation of salycylate content in *Arabidopsis thaliana* by the expression of an engineered bacterial salicylate synthase. *The Plant Journal* 25: 67-77.

Métraux, J.P. (2002). Recent breakthroughs in the study of salicylic acid biosynthesis. *Trends in Plant Science* 7: 332-334.

Métraux, J.P., Nawrath, C., and Genoud, T. (2002). Systemic acquired resistance. *Euphytica* 124: 237-243.

Molina, A., Hunt, M.D., and Ryals, J.A. (1998). Impaired fungicide activity in plants blocked in disease resistance signal transduction. *Plant Cell* 10: 1903-1914.

Ongena, M., Daayf, F., Jacques, P., Thonart, P., Benhamon, N., Paulitz, T.C., and Belanger, R.R. (2000). Systemic induction of phytoalexins in cucumber in response to treatments with fluorescent pseudomonads. *Plant Physiology* 49: 523-530.

Oostendorp, M., Kunz, W., Dietrich, B., and Staub, T. (2001). Induced disease resistance in plants by chemicals. *European Journal of Plant Pathology* 107: 19-28.

Ortega, F., Steiner, U., and Dehn, H.W. (1998). Induced resistance: A tool for fungicide resistance management. *Pesticide Science* 53: 193-196.

Pajot, E., Le Corre, D., and Silue, D. (2001). Phytogard and DL-B-amino butyric acid (BABA) induce resistance to downy mildew *(Bremia lactucae)* in lettuce *(Lactuca sativa* L.). *European Journal of Plant Pathology* 107: 861-869.

Pieterse, C.M.J., Van Pelt, J.A., Ton, J., Parchmann, S., Mueller, M.J., Buchala, A.J., Métraux, J.P., and Van Loon, L.C. (2000). Rhizobacteria-mediated induced systemic resistance (ISR) in *Arabidopsis* requires sensitivity to jasmonate and ethylene butis not accompanied by an increase in their production. *Physiological and Molecular Plant Pathology* 57: 123-134.

Pieterse, C.M.J., Van Pelt, J.A, Van Wees, S.C.M., Ton, J., Leon-Kloosterziel, K.M., Keurentjes, J.J.B., Verhagen, B.W.M., Knoester, M., Van der Sluis, I., Bakker, P.A.H.M.,and Van Loon, L.C. (2001). Rhizobacteria-mediated induced systemic resistance: Triggering, signalling and expression. *European Journal of Plant Pathology* 107: 51-61.

Pieterse, C.M.J., Van Wees, S.C.M., Van Pelt, J.A., Knoester, M., and Laan, R. (1998). A novel signaling pathway controlling induced systemic resistance in *Arabidopsis. The Plant Cell* 10: 1571-1580.

Prats, E., Rubiales, D., and Jorrín, J. (2002) Acibenzolar-*S*-methyl-induced resistance to sunflower rust *(Puccinia helianthi)* is associated with an enhancement of coumarins on foliar surface. *Physiological and Molecular Plant Pathology* 60: 155-162.

Punja, Z.K. (2001). Genetic engineering of plants to enhance resistance to fungal pathogens—A review of progress and future progress. *Canadian Journal of Plant Pathology* 23: 216-235.

Renwick, J.A.A. (1997). Diversity and dynamics of crucifer defenses against adults and larvae of cabbage butterflies. *Recent Advances in Phytochemistry* 30: 57-80.

Ross, A.F. (1961). Systemic acquired resistance induced by localized virus infection in plants. *Virology* 14: 340-358.

Ryals, J., Weymann, K., Lawton, K., Friedrich, L., Ellis, D., Steiner, H-Y., Johnson, J., Delaney, T., Taco, J., Vos, P., and Uknes, S. (1997). The *Arabidopsis* NIM1 protein shows homology to the mammalian transcription factor inhibitor IkB. *The Plant Cell* 9: 425-439.

Sakamoto, K., Tada, Y., Yokezeki, Y., Akagi, H., Hayashi, N., Fujimura, T., and Ichikawa, N. (1999). Chemical induction of disease resistance in rice is correlated with the expression of a gene encoding a nucleotide binding site and a leucine-rich repeat. *Plant Molecular Biology* 40: 847-855.

Stadnik, M.J. and Buchenauer, H. (2000). Inhibition of phenylalanine ammonia-lyase suppresses the resistance induced by benzothiadiazole in wheat to *Blumeria graminis* f. sp. *tritici*. *Physiological and Molecular Plant Pathology* 57: 25-34.

Sticher, L., Mauch-Mani, B., and Métraux, J.P. (1997). Systemic acquired resistance. *Annual Review of Phytopathology* 35: 235-270.

Tally, A., Oostendorp, M., Lawton, K., Staub, T., and Bassi, B. (1999). Commercial development of elicitors of induced resistance to pathogens. In Agrawal, A.A., Tuzun, S., and Bent, E. (Eds.), *Induced Plant Defense Against Pathogens and Herbivores: Biochemistry, Ecology and Agriculture* (pp. 357-369). St. Paul, MN: American Phytopathological Society Press.

Ton, J., Van Pelt, J.A., Van Loon, L.C., and Pieterse, C.M.J. (2002). Differential effectiveness of salicylate-dependent and jasmonate/ethylene-dependent induced resistance in *Arabidopsis*. *Molecular Plant-Microbe Interactions* 15: 27-34.

Van Loon, L.C. (1997). Induced resistance in plants and the role of pathogenesis-related proteins. *European Journal of Plant Pathology* 103: 753-765.

Van Loon, L.C., Bakker, P.A.H.N., and Pieterse, C.M.J. (1998). Systemic resistance induced by rhizosphere bacteria. *Annual Review of Phytopathology* 36: 453-483.

Van Peer, R., Niemann, G.J., and Schippers, B. (1991). Induced resistance and phytoalexin accumulation in biological control of *Fusarium* wilt of carnation by *Pseudomonas* sp. strain WCS417r. *Phytopathology* 81: 728-734.

Van Wees, S.C.M., de Swart, E.A.M., Van Pelt, J.A., Van Loon, L.C., and Pieterse, C.M.J. (2000). Enhancement of induced disease resistance by simultaneous activation of salicylate- and jasmonate-dependent pathways in *Arabidopsis thaliana*. *Proceedings of the National Academy of Sciences, USA* 97: 8711-8716.

Ward, E.W.B. (1984). Suppression of metalaxyl activity by glyphosate: Evidence that host defence mechanisms contribute to metalaxyl inhibition of *Phytophthora megasperma* f. sp. *glycinea* in soybeans. *Physiological Plant Pathology* 25: 381-386.

White, R.F. (1979). Acetylsalicylic acid (aspirin) induces resistance to tobacco mosaic virus in tobacco. *Virology* 99: 410-412.

Wildermuth, M.C., Dewdney, J., Wu, G., and Ausubel, F.M. (2001). Isochorisinate synthase is required to synthesize salicylic acid for plant defence. *Nature* 414: 562-576.

Yoshioka, K., Nakashita, H., Klessig, D.F., and Yamaguchi, I. (2001). Probenzole induces systemic acquired resistance in *Arabidopsis* with a novel type of action. *Plant Journal* 25: 149-157.

Zhang, Y., Fan, W., Kinkema, M., Li, X., and Dong, X. (1999). Interaction of NPR1 with basic leucine zipper transcription factors that bind sequences required for salicylic acid induction of the PR-1 gene. *Proceedings of the National Academy of Sciences, USA,* 96: 6523-6528.

Zhender, G.W., Murphy, J.F., Sikora, E.J., and Kloepper, J.W. (2001). Application of rhizobacteria for induced resistance. *European Journal of Plant Pathology* 107: 39-50.

Chapter 7

Genetic Engineering of Plants to Enhance Resistance to Fungal Pathogens

Zamir K. Punja

INTRODUCTION

From the time domestication of plants for human use first began, fungal diseases have caused major yield losses and have impacted the well-being of humans worldwide. The incorporation of disease-resistance genes into plants has been successfully achieved using conventional breeding methods, and almost every agricultural crop grown today has some form of genetic resistance, generally against a number of diseases. Without these resistance genes, crop productivity and yield would be substantially reduced (Agrios, 1997).

A major area of research in plant biology has been to identify, clone, and characterize the various genes involved in disease resistance. The mechanisms that plants have evolved to respond to pathogen infection have been recently identified, and remarkable progress has been made toward elucidating the multitude of genes involved in these responses. Through the identification of these genes, it is now possible to evaluate their specific roles in the disease-response pathway using transgenic plants developed through genetic engineering techniques. In this chapter, the use of cloned genes (from both plant and nonplant sources) to enhance disease resistance against fungal pathogens in transgenic plants will be reviewed.

The following approaches have been taken to develop disease-resistant transgenic plants.

1. Expression of gene products that are directly toxic to or which reduce growth of the pathogen, including pathogenesis-related (PR) proteins, such as hydrolytic enzymes (chitinases, glucanases) and antifungal proteins (osmotin, thaumatin-like), as well as antimicrobial peptides (thionins, defensins, lectin), ribosome-inactivating proteins, and phytoalexins
2. Expression of gene products that destroy or neutralize a component of the pathogen arsenal, including inhibition of polygalacturonase, oxalic acid, and lipase
3. Expression of gene products that can potentially enhance the structural defenses in the plant, including elevated levels of peroxidase and lignin
4. Expression of gene products that release signals that can regulate plant defenses, including the production of specific elicitors, hydrogen peroxide, salicylic acid, and ethylene
5. Expression of resistance gene (R) products involved in R/Avr interactions and the hypersensitive response (HR)

The choice of genes used to engineer plants to protect against fungal diseases has been based, in part, on the toxicity of the gene product to fungal growth or development in vitro, and to the prominence of the particular gene(s) in a disease-resistance-response pathway. Although some of these genes may normally be expressed relatively late in the response pathway, e.g., after 48 h, the rationale for developing transgenic plants was to achieve early and high expression (overexpression) of these proteins, usually constitutively throughout most of the plant. In other instances, enhancement of protein levels was reasoned to provide a greater inhibitory effect on fungal development than lower naturally occurring or induced levels in the plant. Other genes have been selected for genetic engineering efforts because of their ability to induce an array of naturally occurring defense mechanisms in the plant (Shah et al., 1995; Swords et al., 1997). More recently, the cloning of several R genes has precipitated interest in utilizing these genes to provide broad-spectrum disease resistance. A few genetic engineering approaches have been based on novel approaches of introducing genes from dsRNA entities (viruses) found in fungi (Clausen et al., 2000) and genes (lysozyme) cloned from human tissues (Nakajima et al., 1997; Takaichi and Oeda, 2000) and also originating from a range of microbes (Lorito and Scala, 1999).

GENETIC ENGINEERING APPROACHES

Hydrolytic Enzymes

The most widely used approach has been to overexpress chitinases and glucanases, which belong to the group of pathogenesis-related proteins (Neuhaus, 1999), and which have been shown to exhibit antifungal activity in vitro (Boller, 1993; Yun et al., 1997). Because chitins and glucans are major components of the cell wall of many groups of fungi, the overexpression of these enzymes in plant cells was postulated to cause lysis of hyphae and thereby reduce fungal growth (Mauch and Staehelin, 1989). The specific roles of these hydrolases in resistance to disease have been difficult to prove in nontransgenic plants, since the enzymes are frequently encountered in both resistant and susceptible tissues, and their expression can also be induced by environmental triggers and plant senescence (Punja and Zhang, 1993). However, following expression of different types of chitinases in a range of transgenic plant species, the rate of lesion development and the overall size and number of lesions were observed to be reduced upon challenge with many fungal pathogens (see Table 7.1 at the end of this chapter). This included fungi with a broad host range, such as *Botrytis cinerea* and *Rhizoctonia solani*. However, chitinase overexpression was ineffective against other pathogens, such as *Cercospora nicotianae*, *Colletotrichum lagenarium,* and *Pythium* spp., indicating that differences exist in the sensitivity of fungi to chitinase. Since the characteristics of chitinases from different sources can vary, e.g., in substrate binding specificity, pH optimum, and localization in the cell, leading to differences in antifungal activity (Sela-Buurlage et al., 1993; Van Loon and Van Strien, 1999), appropriate selection of the gene to be used against a targeted pathogen or group of pathogens is required. Although the results from the efforts to date have not been spectacular in terms of the level of disease control, they show that the rate of disease progress and overall disease severity can be significantly reduced. A few transgenic crop species expressing chitinases have been evaluated in field trials, and it was demonstrated that disease incidence was reduced (Howie et al., 1994; Grison et al., 1996; Melchers and Stuiver, 2000).

There are fewer examples of the expression of glucanases in transgenic plants (Table 7.1), but the results have generally been similar to those for chitinase expression. The combined expression of chitinases and glucanases in transgenic carrot, tomato, and tobacco was much more effective in preventing the development of disease caused by a number of pathogens compared to either one alone (Van den Elzen et al., 1993; Zhu et al., 1994; Jongedijk et al., 1995), confirming the synergistic activity of these two enzymes reported from in vitro studies (Sela-Buurlage et al., 1993; Van den Elzen et al., 1993; Melchers and Stuiver, 2000). As a general rule, the deployment of genetic engineering approaches that involve the expression of two or more antifungal gene products in a specific crop should provide more effective and broad-spectrum disease control than the single-gene strategy (Lamb et al., 1992; Cornelissen and Melchers, 1993; Strittmatter and Wegner, 1993; Jach et al., 1995; Shah, 1997; Evans and Greenland, 1998; Salmeron and Vernooij, 1998; Melchers and Stuiver, 2000).

Pathogenesis-Related Proteins

Other PR proteins that exhibit antifungal activity, including osmotin and thaumatin-like protein, and some uncharacterized PR proteins, have also been engineered into crop plants (Table 7.1). Osmotin is a basic 24 kDa protein belonging to the PR-5 family whose members have a high degree of homology to the sweet-tasting protein thaumatin from *Thaumatococcus daniellii* and are produced in plants under different stress conditions (Zhu et al., 1995). The PR-5 proteins induce fungal cell leakiness, presumably through a specific interaction with the plasma membrane that results in the formation of transmembrane pores (Kitajima and Sato, 1999). Osmotin has been shown to have antifungal activity in vitro (Woloshuk et al., 1991; Melchers et al., 1993; Liu et al., 1994), and when tested in combination with chitinase and β-1,3-glucanse, it showed enhanced lytic activity (Lorito et al., 1996). When expressed in transgenic potato, osmotin was shown to delay expression of disease symptoms caused by *Phytophthora infestans* (Table 7.1). Thaumatin-like proteins (TLP) are also expressed in plants in response to a range of stress conditions and have also been demonstrated to have antifungal activity in vitro (Malehorn et al., 1994; Koiwa et al., 1997). Expression of TLP in transgenic plants was reported to delay disease development due to several pathogens, including *Botrytis, Fusarium, Rhizoctonia,* and *Sclerotinia* (Table 7.1). Combinations of PR-5 protein expression with chitinases or glucanases in transgenic plants have not been reported, but it is anticipated that the level of disease reduction achieved would be enhanced.

Antimicrobial Proteins/Peptides/Other Compounds

Defensins and thionins are low molecular weight (around 5 kDa) cysteine-rich peptides (45 to 54 amino acids in length) found in monocotyledenous and dicotyledenous plant species, that were initially derived from seeds, and which have antimicrobial activity (Bohlmann, 1994; Broekaert et al., 1995; Evans and Greenland, 1998). It was proposed that these peptides played a role in protecting seeds from infection by pathogens (Broekaert et al., 1997). Defensins are also found in insects and mammals, where they play an important role in curtailing or limiting microbial attack (Rao, 1995). These peptides may exert antifungal activity by altering membrane permeability and/or inhibiting macromolecule biosynthesis, and thionins may be toxic to plant and animal cell cultures as well (Broekaert et al., 1997). The overexpression of defensins and thionins in transgenic plants was demonstrated to reduce development of several different pathogens, including *Alternaria, Fusarium,* and *Plasmodiophora* (Table 7.1), and provided resistance to *Verticillium* on potato under field conditions (Gao et al., 2000).

Chitin-binding peptides (hevein-type, knottin-type) are 36 to 40 residues in length and have been recovered from the seeds of some plant species. They contain cysteine residues and have been demonstrated to have antifungal activity in vitro (Broekaert et al., 1997). However, expression of *Amaranthus* hevein-type peptide and *Mirabilis* knottin-type peptide in transgenic tobacco did not enhance tolerance to *Alternaria longipes* or *Botrytis cinerea* (De Bolle et al., 1996). It was postulated that the presence of cations, particularly Ca^{2+}, may have inhibited the activity of these peptides in vivo. Modifications to amino acid sequences of peptides may enhance the antifungal activity (Evans and Greenland, 1998).

Ribosome-inactivating proteins (RIP) are plant enzymes that have 28 S rRNA N-glycosidase activity which, depending on their specificity, can inactivate nonspecific or foreign ribosomes, thereby shutting down protein synthesis. The most common cytosolic type I RIP from the endosperm of cereal grains do not act on plant ribosomes but can affect foreign ribosomes, such as those of fungi (Stirpe et al., 1992; Hartley et al., 1996). Expression of barley seed RIP reduced development of *Rhizoctonia solani* in transgenic tobacco (Logemann et al., 1992) but had little effect on *Blumeria graminis* in transgenic wheat (Bieri et al., 2000). In the latter study, the RIP was targeted to the apoplastic space and may have had less activity against development of the intracellular haustoria of the mildew pathogen. It has been demonstrated that combined expression of chitinase and RIP in transgenic tobacco had a more inhibitory effect on *R. solani* development than the individual proteins (Jach et al., 1995). Therefore, dissolution of the fungal cell wall by hydrolytic enzymes should enhance the efficacy of antifungal proteins and peptides in transgenic plants. Human lysozyme has lytic activity against fungi and bacteria, and when expressed in transgenic carrot and tobacco, enhanced resistance to several pathogens, including *Erysiphe* and *Alternaria* (Nakajima et al., 1997; Takaichi and Oeda, 2000). An antimicrobial protein with homology to lipid transfer protein was shown to reduce development of *B. cinerea* when expressed in transgenic geranium (Bi et al., 1999).

Pokeweed *(Pytolacca americana)* antiviral protein with type I RIP activity has been expressed in transgenic tobacco and was shown to reduce development of *R. solani* (Wang et al., 1998). Because of some toxicity to plant cells, nontoxic mutant proteins were derived and their expression in transgenic plants led to the activation of defense-related signaling pathways and PR-protein induction (e.g., chitinase and glucanase), which in turn enhanced plant resistance to infection by *R. solani* (Zoubenko et al., 1997). The induction of defense pathways in transgenic plants using other strategies will be discussed in the following sections.

Antimicrobial peptides have been synthesized in the laboratory to produce smaller (10 to 20 amino acids in length) molecules that have enhanced potency against fungi (Cary et al., 2000). In addition, a synthetic cationic peptide chimera (cecropin-melittin) with broad-spectrum antimicrobial activity has been produced (Osusky et al., 2000). When expressed in transgenic potato and tobacco, these synthetic peptides provided enhanced resistance against a number of fungal pathogens, including species of *Colletotrichum, Fusarium,* and *Phytophthora* (Table 7.1). These peptides may demonstrate lytic activity against fungal hyphae, inhibit cell wall formation, and/or enhance membrane leakage. The ability to create synthetic recombinant and combinatorial variants of peptides that can be rapidly screened in the laboratory should provide additional opportunities to engineer resistance to a range of pathogens simultaneously. Enhancement of the specific activities of antifungal enzymes or the creation of variants with broad activity using directed molecular evolution (DNA shuffling) has also been proposed as a method to enhance the efficacy of transgenic plants in the future (Lassner and Bedbrook, 2001).

Phytoalexins

Phytoalexins are low molecular weight secondary metabolites produced in a broad range of plant species that have been demonstrated to have antimicrobial activity and are induced by pathogen infection and elicitors (Hammerschmidt, 1999; Grayer and Kokubun, 2001). Phytoalexins are synthesized through complex biochemical pathways (Dixon et al., 1996), such as the shikimic acid pathway, and genetic manipulation of these pathways to suppress or enhance phytoalexin production has been difficult to achieve. As with the hydrolytic enzymes, it has not been easy to conclusively demonstrate the role played by phytoalexins in enhancing resistance to disease in many host-pathogen interactions. A mutant of *Arabidopsis* deficient in the production of the indole-type phytoalexin camalexin was shown to be more susceptible to infection by *Alternaria brassicicola* but not to *Botrytis cinerea* (Thomma, Nelissne, et al., 1999). Using transgenic plants, it has been possible to also show that the overexpression of genes encoding certain phytoalexins, such as *trans*-resveratrol and medicarpin, resulted in delayed development of disease and symptom production by a number of pathogens on several plant species (Table 7.1). These studies are encouraging in light of the difficulties of engineering the complex biochemical pathways leading to phytoalexin accumulation in plants (Dixon et al., 1996).

Inhibition of Pathogen Virulence Products

The plant cell wall acts as a barrier to penetration by fungal pathogens, and numerous strategies have evolved among plant pathogens to overcome this (Walton, 1994). These include secretion of a range of plant cell wall–degrading enzymes (depolymerases) and the production of toxins, such as oxalic acid, by fungal pathogens. A large number of genes involved in pathogenicity of fungi have been indentified (Idnurm and Howlett, 2001). Several strategies to engineer resistance against fungal infection have targeted the inactivation of these pathogen virulence products. Polygalacturonase-inhibiting proteins (PGIP) are glycoproteins present in the cell wall of many plants that can inhibit the activity of fungal endopolygalacturonases (Powell et al., 1994; Desiderio et al., 1997; Lorenzo and Ferrari, 2002). The expression of PGIP in transgenic plants produced contrasting results: in transgenic tomato expressing a bean PGIP, resistance to *Fusarium, Botrytis,* or *Alternaria* was not enhanced (Desiderio et al., 1997), while in transgenic tomato expressing a pear PGIP, colonization of leaves and fruits by *Botrytis* was reduced (Powell et al., 2000). In the former study, it was shown that PGIPs from bean differed in their specificity to fungal PG in vitro and the PGIP-1 that was selected for transformation was not inhibitory (Desiderio et al., 1997). Thus, appropriate in vitro screening of PGIPs would be required prior to undertaking transformation experiments. As with the PR proteins and antifungal compounds, disease development was reduced but not totally prevented in the transgenic plants.

Another strategy that could have potential to reduce pathogen infection is immunomodulation or the expression of genes encoding antibodies or antibody fragments in plants (plantibodies) that could bind to pathogen virulence products (De Jaeger et al., 2000; Schillberg et al., 2001). The antibodies can be expressed intercellularly or extracellularly and can bind to and inactivate enzymes, toxins, or other pathogen factors involved in disease development. Currently, no published reports on the expression of antifungal antibodies in transgenic plants have led to a reduction in disease. However, it has been demonstrated that antilipase antibodies inhibited infection of tomato by *B. cinerea* when mixed with spore inoculum by preventing fungal penetration through the cuticle (Comménil et al., 1998). Similarly, infection by *Colletotrichum gloeosporioides* on various fruits was inhibited using polyclonal antibodies that bound to fungal pectate lyase (Wattad et al., 1997). Genetic engineering of antibody expression in plants is extremely challenging technically and the applications to fungal disease control (immunization) have yet to be determined, although success against virus diseases has been reported (De Jaeger et al., 2000).

Production of phytotoxic metabolites, such as mycotoxins and oxalic acid, by fungal pathogens has been shown to facilitate infection of host tissues following cell death. Degradation of these compounds by enzymes expressed in transgenic plants could provide an opportunity to enhance resistance to disease. Expression of a trichothecene-degrading enzyme from *Fusarium sporotrichioides* in transgenic tobacco reduced plant tissue damage and enhanced seedling emergence in the presence of the trichothecene (Muhitch et al., 2000). The effect on pathogen development was not tested. Germinlike oxalate oxidases are stable glycoproteins first discovered in cereals that are present during seed germination and induced in response to fungal infection and abiotic stress (Dumas et al., 1995; Zhang et al., 1995; Berna and Bernier, 1997). Their activity on the substrate oxalic acid results in the production of CO_2 and H_2O_2; the latter can induce defense responses in the plant and enhance strengthening of cell walls (Brisson et al., 1994; Mehdy, 1994). The expression of barley oxalate oxidase in oilseed rape enhanced tolerance to the phytotoxic effects of oxalic acid, although the effect on the target pathogen *Sclerotinia sclerotiorum* was not evaluated (Thompson et al., 1995). In transgenic soybean, a wheat oxalate oxidase provided resistance to *S. sclerotiorum* (Donaldson et al., 2001). Expression of oxalate oxidase in transgenic hybrid poplar enhanced resistance to *Septoria,* whereas oxalate decarboxylase expression enhanced resistance of tomato to *Sclerotinia sclerotiorum* (Table 7.1). These reports indicate that the inactivation of specific pathogen virulence factors, such as toxins, by gene products expressed in transgenic plants has the potential to reduce development of specific fungal pathogens.

Alteration of Structural Components

Lignification of plant cells around sites of infection or lesions has been reported to be a defense response of plants that can potentially slow down pathogen spread (Nicholson and Hammerschmidt, 1992). The enzyme peroxidase is required for the final polymerization of phenolic derivatives into lignin and may also be involved in suberization or wound healing. A decrease in polyphenolic compounds, such as lignin, in potato tubers by redirection of tryptophan in transgenic plants through expression of tryptophan decarboxylase rendered tissues more susceptible to *Phytophthora infestans* (Yao et al., 1995), illustrating the role of phenolic compounds in defense. Reduction of phenylpropanoid metabolism through inhibition of phenylalanine ammonia-lyase activity in transgenic tobacco also rendered tissues more susceptible to *Cercospora nicotianae* (Maher et al., 1994). Overexpression of a cucumber peroxidase gene in transgenic potato, however, did not increase resistance of tissues to infection by *Fusarium* or *Phytophthora,* and lignin levels were not significantly affected, despite elevated peroxidase expression (Ray et al., 1998). It was suggested that peroxidase levels may not have been the limiting step for lignification or that the native peroxidase activity may have been co-suppressed. Overexpression of a tobacco anionic peroxidase gene in tomato did enhance lignin levels, but resistance to fungal pathogens was not enhanced (Lagrimini et al., 1993). Lignin levels were also significantly higher following expression of the H_2O_2-generating enzyme glucose oxidase in transgenic potato (Wu et al., 1997) and by expression of the hormone indole-acetic acid in transgenic tobacco (Sitbon et al., 1999). In the former case, tolerance to several fungal pathogens was enhanced (Table 7.1). Peroxidase overexpression in plants can, however, have negative effects on plant growth and development (Lagrimini et al., 1997), and the results to date indicate that this approach appears to hold less promise for enhancing disease resistance.

A reduction in large callose deposits surrounding haustoria of *Peronospora parasitica* infecting *Arabidopsis thaliana* was indirectly achieved in transgenic plants not accummulating salicylic acid, achieved by expression of the enzyme salicylate hydroxylase (Donofrio and Delaney, 2001). These plants also had reduced expression of the PR-1 gene and exhibited significantly enhanced susceptibility to the pathogen, suggesting that callose deposition during normal defense responses of the plant was influenced by the reduced levels of SA.

ACTIVATION OF PLANT DEFENSE RESPONSES

One activator of host defense responses are elicitor molecules from an invading pathogen. These can trigger a network of signaling pathways that coordinate the defense responses of the plant, including the hypersensitive response, and PR-protein and phytoalexin production (Heath, 2000; McDowell and Dangl, 2000; Shirasu and Schulze-Lefert, 2000). A gene encoding the elicitor cryptogein (a small basic protein, 98 amino acids in length) from the pathogen *Phytophthora cryptogea* was cloned and expressed in transgenic tobacco under control of a pathogen-inducible promoter (Keller et al., 1999). Challenge inoculation with a range of fungi induced the HR as well as several defense genes, and growth of the pathogens was concomitantly restricted (Table 7.1). Resistance to the pathogens was not complete, possibly because of the time needed for production of the transgenic elicitor following initial infection (Keller et al., 1999). Another elicitor, INF1, was shown to act as an avirulence factor in the tobacco-*Phytophthora infestans* interaction and triggered the onset of the HR (Kamoun et al., 1998). Expression of the gene encoding the AVR9 peptide elicitor from *Cladosporium fulvum* in transgenic tomatoes containing the *Cf9* gene resulted in a necrotic defense response (Hammond-Kosack et al., 1994; Honée et al., 1995). The development of lesions resembling the HR induced through expression of a bacterial proton pump gene (bacterio-opsin) from *Halobacterium halobium* activated multiple defense systems in transgenic tobacco plants (Mittler et al., 1995) in the absence of pathogen challenge.

In transgenic potato, expression of bacterio-opsin enhanced resistance to some pathogens but had no effect on others (Abad et al., 1997), whereas in poplar, there was no effect on disease development (Mohamed et al., 2001). Antisense inhibition of catalase, a H_2O_2-degrading enzyme, resulted in development of necrotic lesions and PR-protein accumulation (Takahashi et al., 1997). Although these and other reports indicate that induction of the HR and necrosis, with the resulting activation of general defense pathways, could potentially result in broad-spectrum disease resistance (Bent, 1996; Honée, 1999; Melchers and Stuiver, 2000), the use of such an approach would require tight regulation of the expressed phenotype, in addition to ensuring that no deleterious side effects, such as abnormal or suppressed growth, occurred on the transgenic plants. If successful, the activation of general defense responses in these transgenic plants could provide protection against viral and bacterial pathogens in addition to fungi.

Another activator of defense responses that has been engineered in transgenic plants is hydrogen peroxide (H_2O_2) generated through expression of genes encoding for glucose oxidase (Table 7.1). Hydrogen peroxide has been shown to directly inhibit pathogen growth (Wu et al., 1995) and to induce PR proteins, salicylic acid, and ethylene (Wu et al., 1997; Chamnongpol et al., 1998), as well as phytoalexins (Mehdy, 1994). It is produced during the early oxidative burst in plant cell response to infection (Baker and Orlandi, 1995) and can trigger the HR (Levine et al., 1994; Tenhaken et al., 1995, Neill et al., 2002), strengthen cell walls (Brisson et al., 1994), and enhance lignin formation (Wu et al., 1997). Expression of elevated levels of H_2O_2 in transgenic cotton, tobacco, and potato reduced disease development due to a number of different fungi, including *Rhizoctonia, Verticillium, Phytophthora,* and *Alternaria* (Table 7.1); high levels can, however, be phytotoxic (Murray et al., 1999). In one study, necrotic lesions from the HR enhanced infection by the necrotrophic pathogen *Botrytis cinerea* (Govrin and Levine, 2000), and the characteristic features of such necrotrophic fungi have been reviewed (Mayer et al., 2001). Therefore, the widespread induction of cell death in a transgenic plant to induce disease resistance has to be approached with caution.

Other activators of plant defense responses include signaling molecules such as salicylic acid, ethylene, and jasmonic acid (Yang et al., 1997; Dong, 1998; Reymond and Farmer, 1998; Dempsey et al., 1999). The roles of salicyclic acid as a signal molecule for the activation of plant defense responses to pathogen infection and as an inducer of systemic acquired resistance (SAR) have been extensively studied (Ryals et al., 1996; Stichter et al., 1997; Dempsey et al., 1999; Métraux, 2002). Using transgenic plants, evidence for the role of SA in defense response activation has been obtained. Plants expressing the SA-metabolizing enzyme salicylate hydroxylase, a bacterial protein that converts SA to the inactive form catechol, did not accumulate high levels of SA and had enhanced susceptibility to pathogen infection (Gaffney et al., 1993; Delaney et al., 1994; Donofrio and Delaney, 2001) or had unaltered susceptibility (Yu et al., 1997). A mutant of *Arabidopsis* nonresponsive to induction of SAR showed enhanced susceptibility to fungal infection (Delaney et al. 1995; Donofrio and Delaney, 2001). The overexpression of SA in transgenic tobacco was recently shown to enhance PR-protein production and provide resistance to fungal pathogens (Verberne et al., 2000). Expression of tobacco catalase, an enzyme with SA-binding activity, in transgenic potato enhanced defense gene expression leading to SAR and enhanced tolerance to *P. infestans* (Yu et al., 1999). Overexpression of the *NPR1* gene, which regulates the SA-mediated signal leading to SAR induction, in transgenic *Arabidopsis* increased the level of PR proteins during infection and enhanced resistance to *Peronospora parasitica* (Cao et al., 1998). It was postulated that synergistic interactions between PR proteins and products of other downstream defense-related genes provided the enhanced resistance. In addition, *NPR1* was only activated upon infection or by induction of SAR, avoiding potential side effects on plant growth from constitutive expression. These studies demonstrate that manipulation of salicyclic acid levels in transgenic plants has the potential to lead to enhanced disease resistance by inducing PR-protein expression and other defense gene products.

Ethylene and jasmonic acid appear to be signals used in response of plants to necrotrophic pathogen attack (in contrast to biotrophic infection) and that work independently of, and possibly antagonistic to, SA-mediated responses (Dong, 1998; Thomma, Nelissen, et al., 1999; McDowell and Dangl, 2000; Lee et al., 2001). Mutant or transformed plants nonresponsive to either jasmonate or ethylene were found to be more susceptible to infection by root- and foliar-infecting fungi (Knoester et al., 1998; Staswick et al., 1998; Vijayan et al., 1998; Hoffmann et al., 1999; Thomma, Eggermont, et al., 1999; Geraats et al, 2002), confirming a role for these signals in certain host-pathogen interactions. In contrast, ethylene-insensitive mutants or plants with reduced ethylene production can exhibit reduced disease symptoms, as described for *F. oxysporum* and *V. dahliae* on tomato (Lund et al., 1998, Robison et al., 2001). Genetic engineering efforts to alter ethylene or jasmonate production in plants may, however, result in unpredictable effects on disease response, depending on the pathogen, as well as induce potential side effects in view of the multiple roles played by these signal molecules in plants (O'Donnell et al., 1996; Weiler, 1997; Wilkinson et al., 1997).

Ethylene production and extracellular PR-protein expression were found to be induced by expression of cytokinins in transgenic tomato cells (Bettini et al., 1998). The engineering of hormone biosynthetic gene expression in transgenic plants has been accomplished (Hedden and Phillips, 2000). Whether reduced or elevated levels of hormones, such as auxins, cytokinins, gibberellins, and jasmonate, can lead to the development of transgenic plants with enhanced disease resistance remains to be seen, given their broad range of physiological effects on plant development. Interestingly, inhibition of indole-acetic acid (IAA) production by antisense transformation of the *nitrilase 1* gene in *Arabidopsis* reduced levels of IAA and development of root galls due to *Plasmodiophora brassicae* (Neuhaus et al., 2000). In contrast, overexpression of IAA in tobacco enhanced ethylene production and peroxidase activity and increased lignin content, although the response to disease was not tested (Sitbon et al., 1999). Altered auxin/cytokinin expression has the potential to also affect mycorrhizal colonization of plant roots (Barker and Tagu, 2000).

RESISTANCE GENES (R GENES)

R-gene products may serve as receptors for pathogen avirulence (Avr) factors or recognize the Avr factor indirectly through a co-receptor (Staskawicz et al., 1995). This gene-for-gene interaction triggers one or more signal transduction pathways which in turn activate defense responses in the plant to prevent pathogen growth (Hammond-Kosack and Jones, 1996; De Wit, 1997). These defense responses include the development of the HR, expression of PR proteins, and accumulation of salicylic acid, and can lead to the development of systemic acquired resistance (Ryals et al., 1996; Dempsey et al., 1999; Kombrink and Schmelzer, 2001). Ethylene and jasmonic acid may also be involved in signaling the defense responses in the gene-for-gene interaction (Deikman, 1997; Dong, 1998). Efforts to clone an array of *R* genes involved in fungal disease resistance have met with some success (Bent, 1996; Crute and Pink, 1996; Baker et al. 1997; Hammond-Kosack and Jones, 1997; Ellis and Jones, 1998). The *R*-gene products cloned from tomato, tobacco, rice, flax, *Arabidopsis,* and several other plant species shared one or more similar motifs: a serine/threonine kinase domain, a nucleotide binding site, a leucine zipper, or a leucine-rich repeat region, all of which may contribute to recognition specificity (Shirasu and Schulze-Lefter, 2000; Takken and Joosten, 2000). The *Hm1 R* gene cloned from maize is an exception, as it encodes for an NADPH-dependent reductase that inactivates the potent toxin produced by race 1 strains of *Cochliobolus carbonum* (Johal and Briggs, 1992). Many *R* genes belong to tightly linked multigene families, e.g., *Cf4/9* encoding resistance to *Cladosporium fulvum* mold of tomato (Thomas et al., 1997).

There are several examples of the expression of *R* genes in transgenic plants. The overexpression of the *HRT* gene, which controls the hypersensitive response in *Arabidopsis* to turnip crinkle virus, did not confer enhanced resistance to *Peronospora tabacina* (Cooley et al., 2000). The authors proposed that multiple factors may be involved in determining the resistance response, or that the resistance may be HR independent. Expression of the *Cf-9* gene, which confers resistance in tomato to races of *C. fulvum,* in transgenic tobacco and potato gave rise to the HR when challenged with Avr 9 peptide (Hammond-Kosack et al., 1998), indicating that the *Cf-9* gene product was produced. Expression of the *Cf-g* gene in oilseed rape enhanced resistance to *Leptosphaeria maculans* (Hennin et al., 2001). However, for the *R-gene/Avr mediated disease resistance to be fully expressed,* several additional loci may be required (Hammond-Kosack and Jones, 1996; Baker et al., 1997), in addition to elevated levels of salicylic acid (Delaney et al., 1994). Results to date suggest that the expression of cloned *R* genes in heterologous transgenic plants is unlikely by itself to enhance tolerance to fungal pathogens, due to the complexity of the interacting signaling pathways: A combination of several interacting genes, similar to that for the antifungal proteins, will likely be required. An enhanced understanding of *R*-gene structure and function could, however, make it possible to modify functional domains in the future to tailor *R* genes for use in providing broad-spectrum resistance to diseases in transgenic plants (Bent, 1996; Dempsey et al., 1998). Other potential approaches to the use of *R* genes for engineering disease resistance in plants are discussed by Rommens and Kishore (2000).

CHALLENGES

Besides identifying and cloning potentially useful genes to engineer into plants, the development of transgenic plants with enhanced fungal disease resistance faces additional challenges. Depending on the plant species, transformation frequencies can be as low as 1 to 10 percent, and from hundreds of confirmed transgenic lines, only a few may have appropriate transgene expression levels. Recent advances in plant transformation should provide new opportunities to overcome some of these difficulties (Gelvin, 1998; Hansen and Wright, 1999; Newell, 2000). The positive relationship of high levels of PR proteins and antifungal compounds with enhanced disease resistance in plants has been documented in many but not all cases. However, as indicated previously, there are a number of examples in which transgene products expressed at high levels induced plant cell damage or had other undesirable effects. These include the engineered expression of thionins, ribosome-inactivating proteins, peroxidase, hydrogen peroxide, elicitor molecules, and growth regulators. In most instances, constitutive promoters have been used to achieve high expression levels throughout most tissues of the plant. In crops affected by pathogens that colonize more than one type of organ, e.g., roots and leaves, this is advantageous. In instances where only specific tissues need to be protected, e.g., leaves, fruit, or seed, or where the antifungal compounds need to be expressed at certain targeted sites in the cell, specific promoters would need to be identified (Bushnell et al., 1998; Dahleen et al., 2001). Wound- and pathogen-inducible promoters have also been described that have advantages for engineering specific disease resistance against fungal pathogens (Roby et al., 1990; Strittmatter et al., 1995; Keller et al., 1999) by expressing antifungal compounds only at sites of infection or wounds. Targeting of the engineered protein to the apoplastic space or to the vacuole has been achieved in numerous previous studies and may enhance the antifungal activity, depending on the mode of infection of the pathogen. Future research will require the fine-tuning of engineered gene expression and establishing the optimal expression levels and target site in the cell needed to prevent pathogen infection.

It remains to be demonstrated under field conditions the level to which disease resistance is achieved using transgenic plants and whether it is against a range of phytopathogens or only specific diseases. It is noteworthy that many of the successfully controlled pathogens in laboratory and greenhouse evaluations are those with a wide host range, such as *Rhizoctonia solani* and *Botrytis cinerea,* for which there are few available sources of genetic resistance through conventional breeding in most crops. This is particularly true also for seedling-infecting pathogens, for which there are few examples of genetic resistance in the host. Therefore, genetic engineering of novel disease-resistance traits in crop plants has the potential to provide control of devastating pathogens with reduced fungicide applications. Expression of an antifungal trait throughout the growing season, from seed to harvest, under prolonged disease-conducive conditions, can also provide significant advantages for disease management using this technology, provided there are no other deleterious side effects.

Nontarget effects on other diseases, pests, or beneficial microorganisms will have to be monitored in crop plants engineered to express antifungal or antimicrobial compounds. Although unpredicted beneficial effects against other related fungal pathogens may be a positive aspect, an assessment of the effects on unrelated fungi, viruses, or bacteria may need to be conducted. It is unwieldy for researchers involved in the development of genetically engineered plants to screen against a multitude of diseases or pathogens common to that crop, an approach that may be taken by plant breeders during development of a new variety. The results in Table 7.1 demonstrate the specificity of the evaluation approach used for transgenic plants, which is conducted mostly under axenic or controlled environment conditions, and which infrequently includes more than one pathogen for challenge inoculation. Could the overexpression of antimicrobial compounds in the roots of genetically engineered plants alter their compatibility with mycorrhizae or beneficial endophytic fungi, or to various rhizosphere-colonizing microbes that could inhibit the development of soilborne pathogens? Evaluations of these potential effects have been conducted in very few studies (Vierheilig et al., 1993, 1995; Lottmann et al., 2000; Lukow et al., 2000), and so far, no side effects have been found.

The possibility of selecting pathogen strains with resistance to the engineered trait may be increased with the widespread deployment of transgenic crops expressing specific antimicrobial compounds or that have broad-spectrum disease resistance. Fungal pathogens have demonstrated the capability for rapid change in genetic structure in the face of selection forces, such as highly specific fungicides, major disease-resistance genes, and environmental factors. The selection imposed by antimicrobial proteins, for example, could force the evolution of adaptive strategies in the pathogen to defend against the inhibitory compounds. Such a co-evolution has been proposed for chitinases (Bishop et al., 2000), in which adaptive functional modifications of the enzyme active site have occurred. Similarly, changes in sensitivity of pathogens to antimicrobial proteins overexpressed in transgenic plants could be selected. The use of combined genes that target different sites could reduce the selection pressure imposed on the pathogen. Genetically engineered plants with successfully enhanced disease resistance should not be viewed as a panacea and continual monitoring for any potential unexpected events will be required. Transgenic plants with enhanced disease resistance can become a valuable component of a disease-management program in the future.

REFERENCES

Abad, M.S., Hakimi, S.M., Kaniewski, W.K., Rommens, C.M., Shulaev, V., Lam, E., and Shah, D.M. (1997). Characterization of acquired resistance in lesion-mimic transgenic potato expressing bacterio-opsin. *Molecular Plant-Microbe Interactions* 10: 635-645.

Agrios, G.N. (1997). *Plant pathology,* Fourth edition. San Diego, CA: Academic Press.

Alexander, D., Goodman, R.M., Gut-Rella, M., Glascock, C., Weymann, K., Friedrich, L., Maddox, D., Ahl-Goy, P., Luntz, T., Ward, E., and Ryals, J.A. (1993). Increased tolerance to two oomycete pathogens in transgenic tobacco expressing pathogenesis-related protein 1a. *Proceedings of the National Academy of Sciences, USA* 90: 7327-7331.

Asao, H., Nishizawa, Y., Arai, S., Sato, T., Hirai, M., Yoshida, K., Shinmyo, A., and Hibi, T. (1997). Enhanced resistance against a fungal pathogen *Sphaerotheca humuli* in transgenic strawberry expressing a rice chitinase gene. *Plant Biotechnology* 14: 145-149.

Baker, B., Zambryski, P., Staskawicz, B., and Dinesh-Kumar, S.P. (1997). Signalling in plant-microbe interactions. *Science* 276: 726-733.

Baker, C.J. and Orlandi, E.W. (1995). Active oxygen in plant pathogenesis. *Annual Review of Phytopathology* 33: 299-322.

Banzat, N., Latorse, M.-P., Bulet, P., Francois, E., Derpierre, C., and Dubald, M. (2002). Expression of insect cystein-rich antifungal peptides in transgenic tobacco enhances resistance to a fungal disease. *Plant Science* 162: 995-1006.

Barker, S.J. and Tagu, D. (2000). The roles of auxins and cytokinins in mycorrhizal symbioses. *Journal of Plant Growth Regulators* 19: 144-154.

Bent, A. (1996). Plant disease resistance genes: Function meets structure. *Plant Cell* 8: 1757-1771.

Berna, A. and Bernier, F. (1997). Regulated expression of a wheat germin gene in tobacco: Oxalate oxidase activity and apoplastic localization of the heterologous protein. *Plant Molecular Biology* 33: 417-429.

Berrocal-Lobo, M., Molina, A., and Solano, R. (2002). Constitutive expression of ethylene-response-factor 1 in *Arabidopsis* confers resistance to several necrotrophic fungi. *The Plant Journal* 29: 23-32.

Bettini, P., Cosi, E., Pellegrini, M.G., Turbanti, L., Vendramin, G.G., and Buiatti, M. (1998). Modification of competence for in vitro response to *Fusarium oxysporum* in tomato cells: III. PR-protein gene expression and ethylene evolution in tomato cell lines transgenic for phytohormone-related bacterial genes. *Theoretical and Applied Genetics* 97: 575-583.

Bi, Y.-M., Cammue, B.P.A., Goodwin, P.H., Krishna Raj, S., and Saxena, P.K. (1999). Resistance of *Botrytis cinerea* in scented geranium transformed with a gene encoding the antimicrobial protein Ace-AmP1. *Plant Cell Reports* 18: 835-840.

Bieri, S., Potrykus, I., and Fütterer, J. (2000). Expression of active barley seed ribosome-inactivating protein in transgenic wheat. *Theoretical and Applied Genetics* 100: 755-763.

Bishop, J.G., Dean, A.M., and Mitchell-Olds, T. (2000). Rapid evolution in plant chitinases: Molecular targets of selection in plant-pathogen coevolution. *Proceedings of the National Academy of Sciences, USA* 97: 5322-5327.

Bliffeld, M., Mundy, J., Potrykus, I., and Fütterer, J. (1999). Genetic engineering of wheat for increased resistance to powdery mildew disease. *Theoretical and Applied Genetics* 98: 1079-1086.

Bohlmann, H. (1994). The role of thionins in plant protection. *Critical Reviews in Plant Science* 13: 1-16.

Bolar, J.P., Norelli, J.L., Harman, G.E., Brown, S.K., and Aldwinckle, H.S. (2001). Synergistic activity of endochitinase and exochitinase from *Trichoderma atroviride (T. harzianum)* against the pathogenic fungus *(Venturia inaequalis)* in transgenic apple plants. *Transgenic Research* 10: 533-543.

Bolar, J.P., Norelli, J.L., Wong, K.-W., Hayes, C.K., Harman, G.E., and Aldwinckle, H.S. (2000). Expression of endochitinase from *Trichoderma harzianum* in transgenic apple increases resistance to apple scab and reduces vigor. *Phytopathology* 90: 72-77.

Boller, T. (1993). Antimicrobial functions of the plant hydrolases, chitinases and β-1,3-glucanases. In Fritig, B. and Legrand, M. (Eds.), *Mechanisms of Plant Defense Responses* (pp. 391-400). Dordrecht, the Netherlands: Kluwer Academic Press.

Brandwagt, B.F., Kneppers, T.J.A., Nijkamp, H.J.J., and Hille, J. (2002). Overexpression of the tomato *Asc-1* gene mediates high insensitivity to *AAL* toxin and fumonisin B_1 in tomato hairy roots and confers resistance to *Alternaria alternata* f. sp. *lycopersici* in *Nicotiana umbratica* plants. *Molecular Plant-Microbe Interactions* 15: 35-42.

Brisson, L.F., Tenhaken, R., and Lamb, C.J. (1994). Function of oxidative cross-linking of cell wall structural proteins in plant disease resistance. *Plant Cell* 6: 1703-1712.

Broekaert, W.F., Cammue, B.P.A., De Bolle, M.F.C., Thevissen, K., De Samblanx, G.W., and Osborn, R.W. (1997). Antimicrobial peptides from plants. *Critical Reviews in Plant Science* 16: 297-323.

Broekaert, W.F., Terras, F.R.G., Cammue, B.P.A., and Osborn, R.W. (1995). Plant defensins: Novel antimicrobial peptides as components of the host defense system. *Plant Physiology* 108: 1353-1358.

Broglie, K., Chet, I., Holliday, M., Cressman, R., Biddle, P. Knowlton, S., Mauvais, C.J., and Broglie, R. (1991). Transgenic plants with enhanced resistance to the fungal pathogen *Rhizoctonia solani*. *Science* 254: 1194-1197.

Broglie, R., Broglie, K., Roby, D., and Chet, I. (1993). Production of transgenic plants with enhanced resistance to microbial pathogens. In Kung S.-D. and Wu, R. (Eds.), *Transgenic Plants,* Volume 1 (pp. 265-276). New York: Academic Press.

Bushnell, W.R., Somers, D.A., Giroux, R.W., Szabo, L.J., and Zeyen, R.J. (1998). Genetic engineering of disease resistance in cereals. *Canadian Journal of Plant Pathology* 20: 137-149.

Cao, H., Li, X., and Dong, X. (1998). Generation of broad-spectrum disease resistance by overexpression of an essential regulatory gene in systemic acquired resistance. *Proceedings of the National Academy of Sciences, USA* 95: 6531-6536.

Cary, J.W., Rajasekaran, K., Jaynes, J.M., and Cleveland, T.E. (2000). Transgenic expression of a gene encoding a synthetic antimicrobial peptide results in inhibition of fungal growth in vitro and in planta. *Plant Science* 154: 171-181.

Chai, B., Magbool, S.B., Hajela, R.K., Green, D., Vargas, J.M. Jr., Warkentin, D., Sabzikar, R., and Sticklen, M.B. (2002). Cloning of a chitinase-like cDNA *(hs2),* its transfer to creeping bentgrass (*Agrostis palustris* Huds.) and development of brown patch *(Rhizoctonia solani)* disease resistant transgenic lines. *Plant Science* 163: 183-193.

Chamnongpol, S., Willekens, H., Moeder, W., Langebartels, C., Sandermann, H., Jr., Van Montagu, M., Inzé, D., and Van Camp, W. (1998). Defense activation and enhanced pathogen tolerance induced by H_2O_2 in transgenic tobacco. *Proceedings of the National Academy of Sciences, USA* 95: 5818-5823.

Chang, M.-M., Chiang, C.C., Martin, M.W., and Hadwiger, L.A. (1993). Expression of a pea disease resistance response gene in the potato cultivar Shepody. *American Potato Journal* 70: 635-647.

Chen, W.P., Chen, P.D., Liu, D.J., Kynast, R., Friebe, B., Velazhahan, R., Muthukrishnan, S., and Gill, B.S. (1999). Development of wheat scab symptoms is delayed in transgenic wheat plants that constitutively express a rice thaumatin-like protein gene. *Theoretical and Applied Genetics* 99: 755-760.

Chen, W.P. and Punja, Z.K. (2002a). *Agrobacterium*-mediated transformation of American ginseng with a rice chitinase gene. *Plant Cell Reports* 20: 1039-1045.

Chen, W.P. and Punja, Z.K. (2002b). Transgenic herbicide- and disease-tolerant carrot (*Daucus carota* L.) plants obtained through *Agrobacterium*-mediated transformation. *Plant Cell Reports* 20: 929-935.

Chong, D.K.X. and Langridge, W.H.R. (2000). Expression of full-length bioactive antimicrobial human lactoferrin in potato plants. *Transgenic Research* 9: 71-78.

Clausen, M., Kräuter, R., Schachermayr, G., Potrykus, I., and Sautter, C. (2000). Antifungal activity of a virally encoded gene in transgenic wheat. *Nature Biotechnology* 18: 446-449.

Comménil, P., Belingheri, L., and Dehorter, B. (1998). Antilipase antibodies prevent infection of tomato leaves by *Botrytis cinerea. Physiological and Molecular Plant Pathology* 52: 1-14.

Constabel, P.C., Bertrand, C., and Brisson, N. (1993). Transgenic potato plants overexpressing the pathogenesis-related STH-2 gene show unaltered susceptibility to *Phytophthora infestans* and potato virus X. *Plant Molecular Biology* 22: 775-782.

Cooley, M.B., Pathirana, S., Wu, H.-J., Kachroo, P., and Klessig, D.F. (2000). Members of the *Arabidopsis HRT/RPP8* family of resistance genes confer resistance to both viral and oomycete pathogens. *Plant Cell* 12: 663-676.

Cornelissen, B.J.C. and Melchers, L.S. (1993). Strategies for control of fungal diseases with transgenic plants. *Plant Physiology* 101: 709-712.

Coutos-Thevenot, P., Poinssot, B., Boromelli, A., Yean, H., Breda, C., Buffard, D., Esnault, R., Hain, R., and Boulay, M. (2001). In vitro tolerance to *Botrytis cinerea* of grapevine 41B rootstock in transgenic plants expressing the stilbene synthase *Vst1* gene under the control of a pathogen-inducible PR10 promoter. *Journal of Experimental Botany* 52: 901-910.

Crute, I.R. and Pink, D.A.C. (1996). Genetics and utilization of pathogen resistance in plants. *Plant Cell* 8: 1747-1755.

Dahleen, L.S., Okubara, P.A., and Blechl, A.E. (2001). Transgenic approaches to combat *Fusarium* head blight in wheat and barley. *Crop Science* 41: 628-637.

Dai, Z., Hooker, B.S., Anderson, D.B., and Thomas, S.R. (2000). Expression of *Acidothermus cellulolyticus* endoglucanase E1 in transgenic tobacco: Biochemical characteristics and physiological effects. *Transgenic Research* 9: 43-54.

Datta, K., Koukolíková-Nicola, Z., Baisakh, N., Oliva, N., and Datta, S.K. (2000). *Agrobacterium*-mediated engineering for sheath blight resistance of indica rice cultivars from different ecosystems. *Theoretical and Applied Genetics* 100: 832-839.

Datta, K., Tu, J., Oliva, N., Ona, I., Velazhahan, R., Mew, T.W., Muthukrishnan, S., and Datta, S.K. (2001). Enhanced resistance to sheath blight by constitutive expression of infection-related rice chitinase in transgenic elite indica rice cultivars. *Plant Science* 160: 405-414.

Datta, K., Velazhahan, R., Oliva, N., Ona, I., Mew, T., Khush, G.S., Muthukrishnan, S., and Datta, S.K. (1999). Overexpression of cloned rice thaumatin-like protein (PR-5) in transgenic rice plants enhances environmental-friendly resistance to *Rhizoctonia solani* causing sheath blight disease. *Theoretical and Applied Genetics* 98: 1138-1145.

De Bolle, M.F.C., Osborn, R.W., Goderis, I.J., Noe, L., Acland, D., Hart, C.A., Torrekens, S., Van Leuven, F., and Broekaert, W.F. (1996). Antimicrobial peptides from *Mirabilis jalapa* and *Amaranthus caudatus:* Expression processing, localization and biological activity in transgenic tobacco. *Plant Molecular Biology* 31: 993-1008.

De Jaeger, G., De Wilde, C., Eeckhout, D., Fiers, E., and Depicker, A. (2000). The plantibody approach: Expression of antibody genes in plants to modulate plant metabolism or to obtain pathogen resistance. *Plant Molecular Biology* 43: 419-428.

De Wit, P.J.G.M. (1997). Pathogen avirulence and plant resistance: A key role for recognition. *Trends in Plant Science* 2: 452-458.

DeGray, G., Rajasekaran, K., Smith, F., Sanford, J., and Daniell, H. (2001). Expression of an antimicrobial peptide via the chloroplast genome to control phytopathogenic bacteria and fungi. *Plant Physiology* 127: 852-862.

Deikman, J. (1997). Molecular mechanism of ethylene regulation of gene transcription. *Physiologica Plantarum* 100: 561-566.

Delaney, T.P., Friedrich, L., and Ryals, J.A. (1995). *Arabidopsis* signal transduction mutant defective in chemically and biologically induced disease resistance. *Proceedings of the National Academy of Sciences, USA* 92: 6602-6606.

Delaney, T.P., Uknes, S., Vernooij, B., Friedrich, L., Weymann, K., Negrotto, D., Gaffney, T., Gut-Rella, M., Kessman, H., Ward, E., and Ryals, J. (1994). A central role of salicylic acid in plant disease resistance. *Science* 266: 1247-1250.

Dempsey, D.A., Shah, J., and Klessig, D.F. (1999). Salicylic acid and disease resistance in plants. *Critical Reviews in Plant Science* 18: 547-575.

Dempsey, D.A., Silva, H., and Klessig, D.F. (1998). Engineering disease and pest resistance in plants. *Trends in Microbiology* 54: 54-61.

Desiderio, A., Aracri, B., Leckie, F., Mattei, B., Salvi, G., Tigelaar, H., Van Roekel, J.S.C., Baulcombe, D.C., Melchers, L.S., De Lorenzo, G., and Cervone, F. (1997). Polygalacturonase-inhibiting proteins (PGIPs) with different specificities are expressed in *Phaseolus vulgaris*. *Molecular Plant-Microbe Interactions* 10: 852-860.

Dixon, R.A., Lamb, C.J., Masoud, S., Sewalt, V.J.H., and Paiva, N.L. (1996). Metabolic engineering: Prospects for crop improvement through the genetic manipulation of phenylpropanoid biosynthesis and defense responses—A review. *Gene* 179: 61-71.

Does, M.P., Houterman, P.M., Dekker, H.L., and Cornelissen, B.J.C. (1999). Processing, targeting, and antifungal activity of stinging nettle agglutin in transgenic tobacco. *Plant Physiology* 120: 421-431.

Donaldson, P.A., Anderson, T., Lane, B.G., Davidson, A.L., and Simmonds, D.H. (2001). Soybean plants expressing an active oligomeric oxalate oxidase from the wheat *gf-2.8* (germin) gene are resistant to the oxalate-secreting pathogen *Sclerotinia sclerotiorum*. *Physiological and Molecular Plant Patholoty* 59: 297-307.

Dong, X. (1998). SA, JA, ethylene and disease resistance in plants. *Current Opinions in Plant Biology* 1: 316-323.

Donofrio, N.M. and Delaney, T.P. (2001). Abnormal callose response phenotype and hypersusceptibility to *Peronospora parasitica* in defense-compromised *Arabidopsis nim 1-1* and salicylate hydroxylase-expressing plants. *Molecular Plant-Microbe Interactions* 14: 439-450.

Dumas, B., Freyssinet, G., and Pallett, K.E. (1995). Tissue-specific expression of germin-like oxalate oxidase during development and fungal infection of barley seedlings. *Plant Physiology* 107: 1091-1096.

El Quakfaoui, S., Potvin, C., Brzezinksi, R., and Asselin, A. (1995). A *Streptomyces* chitosanase is active in transgenic tobacco. *Plant Cell Reports* 15: 222-226.

Ellis, J. and Jones, D. (1998). Structure and function of proteins controlling strain-specific pathogen resistance in plants. *Current Opinion in Plant Science* 1: 288-293.

Epple, P., Apel, K., and Bohlmann, H. (1997). Overexpression of an endogenous thionin enhances resistance of *Arabidopsis* against *Fusarium oxysporum*. *Plant Cell* 9: 509-520.

Evans, I.J. and Greenland, A.J. (1998). Transgenic approaches to disease protection: Applications of antifungal proteins. *Pesticide Science* 54: 353-359.

Fettig, S. and Hess, D. (1999). Expression of a chimeric stilbene synthase gene in transgenic wheat lines. *Transgenic Research* 8: 179-189.

Gaffney, T., Friedrich, L., Vernooij, B., Negrotto, D., Nye, G., Uknes, S., Ward, E., Kessmann, H., and Ryals, J. (1993). Requirement of salicylic acid for the induction of systemic acquired resistance. *Science* 261: 754-756.

Gao, A.-G., Hakimi, S.M., Mittanck, C.A., Wu, Y., Woerner, B.M., Stark, D.M., Shah, D.M., Liang, J., and Rommens, C.M.T. (2000). Fungal pathogen protection in potato by expression of a plant defensin peptide. *Nature Biotechnology* 18: 1307-1310.

Gelvin, S.B. (1998). The introduction and expression of transgenes in plants. *Current Opinion in Biotechnology* 9: 227-232.

Geraats, B.P.J., Bakker, P.A.H.M., and Van Loon, L.C. (2002). Ethylene insensitivity impairs resistance to soilborne pathogens in tobacco and *Arabidopsis thaliana*. *Molecular Plant-Microbe Interactions* 15: 1048-1085.

Govrin, E.M. and Levine, A. (2000). The hypersensitive response facilitates plant infection by the necrotrophic pathogen *Botrytis cinerea*. *Current Biology* 10: 751-757.

Grayer, R.J. and Kokubun, T. (2001). Plant-fungal interactions: The search for phytoalexins and other antifungal compounds from higher plants. *Phytochemistry* 56: 253-263.

Grison, R., Grezes-Besset, B., Scheider, M., Lucante, N., Olsen, L., Leguay, J.-L., and Toppan, A. (1996). Field tolerance to fungal pathogens of *Brassica napus* constitutively expressing a chimeric chitinase gene. *Nature Biotechnology* 14: 643-646.

Hain, R., Reif, H.-J., Krause, E., Langebartels, R., Kindl, H., Vornam, B., Wiese, W., Schmeltzer, E., Schreier, P.H., Stöker, R.H., and Stenzel, K. (1993). Disease resistance results from foreign phytoalexin expression in a novel plant. *Nature* 361: 153-156.

Hammerschmidt, R. (1999). Phytoalexins: What have we learned after 60 years? *Annual Review of Phytopathology* 37: 285-306.

Hammond-Kosack, K.E., Harrison, K., and Jones, J.D.G. (1994). Developmentally regulated cell death on expression of the fungal avirulence gene *Avr 9* in tomato seedlings carrying the disease resistance gene *Cf-9*. *Proceedings of the National Academy of Sciences, USA* 91: 10444-10449.

Hammond-Kosack, K.E. and Jones, J.D.G. (1996). Resistance gene-dependent plant defense responses. *Plant Cell* 8: 1773-1791.

Hammond-Kosack, K.E. and Jones, J.D.G. (1997). Plant disease resistance genes. *Annual Review of Plant Physiology and Plant Molecular Biology* 48: 575-607.

Hammond-Kosack, K.E., Tang, S., Harrison, K., and Jones, J.D.G. (1998). The tomato *Cf-9* disease resistance gene functions in tobacco and potato to confer responsiveness to the fungal avirulence gene product Avr 9. *Plant Cell* 10: 1251-1266.

Hansen, G. and Wright, M.S. (1999). Recent advances in the transformation of plants. *Trends in Plant Science* 4: 226-231.

Harris, L.J. and Gleddie, S.C. (2001). A modified *Rpl3* gene from rice confers tolerance of the *Fusarium graminearum* mycotoxin deoxynivalenol to transgenic tobacco. *Physiological and Molecular Plant Pathology* 58: 173-181.

Hartley, M.R., Chaddock, J.A., and Bonness, M.S. (1996). The structure and function of ribosome inactivating proteins. *Trends in Plant Science* 1: 254-260.

He, X.-Z. and Dixon, R.A. (2000). Genetic manipulation of isoflavone 7-*O*-methyltransferase enhances biosynthesis of 4'-*O*-methylated isoflavonoid phytoalexins and disease resistance in alfalfa. *Plant Cell* 12: 1689-1702.

Heath, M.C. (2000). Hypersensitive response-related death. *Plant Molecular Biology* 44: 321-334.

Hedden, P. and Phillips, A.L. (2000). Manipulation of hormone biosynthetic genes in transgenic plants. *Current Opinion in Biotechnology* 11: 130-137.

Hennin, C., Hofte, M., and Diederichsen, E. (2001). Functional expression of *Cf9* and *Avr9* genes in *Brassica napus* induces enhanced resistance to *Leptosphaeria maculans*. *Molecular Plant-Microbe Interactions* 14: 1075-1085.

Hipskind, J.D. and Paiva, N.L. (2000). Constitutive accumulation of a resveratrol-glucoside in transgenic alfalfa increases resistance to *Phoma medicaginis*. *Molecular Plant-Microbe Interactions* 13: 551-562.

Hoffman, T., Schmidt, J.S., Zheng, X., and Bent, A. (1999). Isolation of ethylene-insensitive soybean mutants that are altered in pathogen susceptibility and gene-for-gene disease resistance. *Plant Physiology* 119: 935-949.

Holtorf, S., Ludwig-Müller, J., Apel, K., and Bohlmann, H. (1998). High-level expression of a viscotoxin in *Arabidopsis thaliana* gives enhanced resistance against *Plasmodiophora brassicae*. *Plant Molecular Biology* 36: 673-680.

Honée, G. (1999). Engineered resistance against fungal pathogens. *European Journal of Plant Pathology* 105: 319-326.

Honée, G., Melchers, L.S., Vleeshouwers, V.G.A.A., van Roekel, J.S.C., and de Wit, P.J.G.M. (1995). Production of the AVR9 elicitor from the fungal pathogen *Cladosporium fulvum* in transgenic tobacco and tomato plants. *Plant Molecular Biology* 29: 909-920.

Howie, W., Joe, L., Newbigin, E., Suslow, T., and Dunsmuir, P. (1994). Transgenic tobacco plants which express the *chiA* gene from *Serratia marcescens* have enhanced tolerance to *Rhizoctonia solani*. *Transgenic Research* 3: 90-98.

Idnurm, A. and Howlett, B.J. (2001). Pathogenicity genes of phytopathogenic fungi. *Molecular Plant Pathology* 2: 241-255.

Jach, G., Görnhardt, B., Mundy, J., Logemann, J., Pinsdorf, E., Leah, R., Schell, J., and Mass, C. (1995). Enhanced quantitative resistance against fungal disease by combinatorial expression of different barley antifungal proteins in transgenic tobacco. *Plant Journal* 8: 97-109.

Jach, G., Logemann, S., Wolf, G., Oppenheim, A., Chet, I., Schell, J., and Logemann, J. (1992). Expression of a bacterial chitinase leads to improved resistance of transgenic tobacco plants against fungal infection. *Biopractice* 1: 33-40.

Johal, G.S. and Briggs, S.P. (1992). Reductase activity encoded by the *HM1* disease resistance gene in maize. *Science* 258: 985-987.

Jongedijk, E., Tigelaar, H., van Roekel, J.S.C., Bres-Vloemans, S.A., Dekker, I., van den Elzen, P.J.M., Cornelissen, B.J.C., and Melchers, L.S. (1995). Synergistic activity of chitinases and β-1,3-glucanases enhances fungal resistance in transgenic tomato plants. *Euphytica* 85: 173-180.

Kamoun, S., van West, P., Vleeshouwers, V.G.A.A., de Groot, K.E., and Govers, F. (1998). Resistance of *Nicotiana benthamiana* to *Phytophthora infestans* is mediated by the recognition of the elicitor protein INF 1. *Plant Cell* 10: 1413-1425.

Kanrar, S., Venkateswari, J.C., Kirti, P.B., and Chopra, V.L. (2002). Transgenic expression of hevein, the rubber tree lectin, in Indian mustard confers protection against *Alternaria brassicae*. *Plant Science* 162: 441-448.

Kazan, K., Rusu, A., Marcus, J.P., Goulter, K.C., and Manners, J.M. (2002). Enhanced quantitative resistance to *Leptosphaeria maculans* conferred by expression of a novel antimicrobial peptide in canola (*Brassica napus* L.). *Molecular Breeding* 10: 63-70.

Keller, H., Pamboukdjian, N., Ponchet, M., Poupet, A., Delon, R., Verrier, J.-L., Roby, D., and Ricci, P. (1999). Pathogen-induced elicitin production in transgenic tobacco generates a hypersensitive response and nonspecific disease resistance. *Plant Cell* 11: 223-235.

Kellmann, J.-W., Kleinow, T., Engelhardt, K., Philipp, C., Wegener, D., Schell, J., and Schreier, P.H. (1996). Characterization of two class II chitinase genes from peanut and expression studies in transgenic tobacco plants. *Plant Molecular Biology* 30: 351-358.

Kesarwani, M., Azam, M., Natarajan, K., Mehta, A., and Datta, A. (2000). Oxalate decarboxylase from *Collybia velutipes:* Molecular cloning and its overexpression to confer resistance to fungal infection in transgenic tobacco and tomato. *Journal of Biological Chemistry* 275: 7230-7238.

Kikkert, J.R., Ali, G.S., Wallace, P.G., Reisch, B., and Reustle, G.M. (2000). Expression of a fungal chitinase in *Vitis vinifera* L. 'Merlot' and 'Chardonnay' plants produced by biolistic transformation. *Acta Horticulture* 528: 297-303.

Kim, J.-K., Duan, X., Wu, R., Seok, S.J., Boston, R.S., Jang, I.-C., Eun, M.-Y., and Nahm, B.H. (1999). Molecular and genetic analysis of transgenic rice plants expressing the maize ribosome-inactivating protein b-32 gene and the herbicide resistance *bar* gene. *Molecular Breeding* 5: 85-94.

Kitajima, S. and Sato, F. (1999). Plant pathogenesis-related proteins: Molecular mechanisms of gene expression and protein function. *Journal of Biochemistry* 125: 1-8.

Knoester, M., Van Loon, L.C., Van den Heuvel, J., Hennig, J., Bol, J.F., and Linthorst, H.J.M. (1998). Ethylene-insensitive tobacco lacks nonhost resistance against soil-borne fungi. *Proceedings of the National Academy of Sciences, USA* 95: 1933-1937.

Koiwa, H., Kato, H., Nakatsu, T., Oda, J., Yamada, Y., and Sato, F. (1997). Purification and characterization of tobacco pathogenesis-related protein PR-5d, an antifungal thaumatin-like protein. *Plant Cell Physiology* 38: 783-791.

Kombrink, E. and Schmelzer, E. (2001). The hypersensitive response and its role in local and systemic disease resistance. *European Journal of Plant Pathology* 107: 69-78.

Krishnamurthy, K., Balconi, C., Sherwood, J.E., and Giroux, M.J. (2001). Wheat puroindolines enhance fungal disease resistance in transgenic rice. *Molecular Plant-Microbe Interactions* 14: 1255-1260.

Krishnaveni, S., Jeoung, J.M., Muthukrishnan, S., and Liang, G.H. (2001). Transgenic sorghum plants constitutively expressing a rice chitinase gene show improved resistance to stalk rot. *Journal of Genetics and Breeding* 55: 151-158.

Lagrimini, L.M., Joly, R.J., Dunlap, J.R., and Liu, T.-T.Y. (1997). The consequence of peroxidase overexpression in transgenic plants on root growth and development. *Plant Molecular Biology* 33: 887-895.

Lagrimini, L.M., Vaughn, J., Erb, W.A., and Miller, S.A. (1993). Peroxidase overproduction in tomato: Wound-induced polyphenol deposition and disease resistance. *HortScience* 28: 218-221.

Lamb, C.J., Ryals, J.A., Ward, E.R., and Dixon, R.A. (1992). Emerging strategies for enhancing crop resistance to microbial pathogens. *Bio/Technology* 10: 1436-1445.

Lassner, M. and Bedbrook, J. (2001). Directed molecular evolution in plant improvement. *Current Opinions in Plant Biology* 4: 152-156.

Leckband, G. and Lörz, H. (1998). Transformation and expression of a stilbene synthase gene of *Vitis vinifera* L. in barley and wheat for increased fungal resistance. *Theoretical and Applied Genetics* 96: 1004-1012.

Lee, M.-W., Qi, M., and Yang, Y. (2001). A novel jasmonic acid-inducible rice *myb* gene associates with fungal infection and host cell death. *Molecular Plant-Microbe Interactions* 14: 527-535.

Lee, Y.H., Yoon, I.S., Suh, S.C., and Kim, H.I. (2002). Enhanced disease resistance in transgenic cabbage and tobacco expressing a glucose oxidase gene from *Aspergillus niger*. *Plant Cell Reports* 20: 857-863.

Levine, A., Tenhaken, R., Dixon, R., and Lamb, C. (1994). H_2O_2 from the oxidative burst orchestrates the plant hypersensitive disease resistance response. *Cell* 79: 583-593.

Li, Q., Lawrence, C.B., Xiang, H.-Y, Babbitt, R.A., Bass, W.T., Maiti, I.B., and Everett, N.P. (2001). Enhanced disease resistance conferred by expression of an antimicrobial magainin analog in transgenic tobacco. *Planta* 212: 635-639.

Liang, H., Catranis, C.M., Maynard, C.A., and Powell, W.A. (2002). Enhanced resistance to the poplar fungal pathogen, *Septoria musiva,* in hybrid poplar clones transformed with genes encoding antimicrobial peptides. *Biotechnology Letters* 24: 383-389.

Liang, H., Maynard, C.A., Allen, R.D., and Powell, W.A. (2001). Increased *Septoria musiva* resistance in transgenic hybrid poplar leaves expressing a wheat oxalate oxidase gene. *Plant Molecular Biology* 45: 619-629.

Lin, W., Anuratha, C.S., Datta, K., Potrykus, I., Muthukrishnan, S., and Datta, S.K. (1995). Genetic engineering of rice for resistance to sheath blight. *Bio/Technology* 13: 686-691.

Liu, D., Raghothama, K.G., Hasegawa, P.M., and Bressan, R. (1994). Osmotin overexpression in potato delays development of disease symptoms. *Proceedings of the National Academy of Sciences, USA* 91: 1888-1892.

Logemann, J., Jach, G., Tommerup, H., Mundy, J., and Schell, J. (1992). Expression of a barley ribosome-inactivating protein leads to increased fungal protection in transgenic tobacco plants. *Bio/Technology* 10: 305-308.

Lorenzo, G.D. and Ferrari, S. (2002). Polygalacturonase-inhibiting proteins in defense against phytopathogenic fungi. *Current Opinion in Plant Biology* 5: 298-299.

Lorito, M. and Scala, F. (1999). Microbial genes expressed in transgenic plants to improve disease resistance. *Journal of Plant Pathology* 81: 73-88.

Lorito, M., Woo, S.L., D'Ambrosio, M., Harman, G.E., Hayes, C.K., Kubicek, C.P., and Scala, F. (1996). Synergistic interaction between cell wall degrading enzymes and membrane affecting compounds. *Molecular Plant-Microbe Interactions* 9: 206-213.

Lorito, M., Woo, S.L., Fernandez, I.G., Colucci, G., Harman, G.E., Pintor-Toro, J.A., Filippone, E., Muccifora, S., Lawrence, C.B., Zoina, A., Tuzun, S., and Scala, F. (1998). Genes from mycoparasitic fungi as a source for improving plant resistance to fungal pathogens. *Proceedings of the National Academy of Sciences, USA* 95: 7860-7865.

Lottman, J., Heuer, H., de Vries, J., Mahn, A., Düring, K. Wackernagel, W., Smalla, K., and Berg, G. (2000). Establishment of introduced antagonistic bacteria in the rhizosphere of transgenic potatoes and their effect on the bacterial community. *FEMS Microbiology and Ecology* 33: 41-49.

Lukow, T., Dunfield, P.F., and Liesack, W. (2000). Use of the T-RFLP technique to assess spatial and temporal changes in the bacterial community structure within an agricultural soil planted with transgenic and non-transgenic potato plants. *FEMS Microbiology and Ecology* 32: 241-247.

Lund, S.T., Stall, R.E., and Klee, H.J. (1998). Ethylene regulates the susceptible response to pathogen infection in tomato. *Plant Cell* 10: 371-382.

Lusso, M. and Kuc, J. (1996). The effect of sense and antisense expression of the *PR-N* gene for β-1,3-glucanase on disease resistance of tobacco to fungi and viruses. *Physiological and Molecular Plant Pathology* 49: 267-283.

Maddaloni, M., Forlani, F., Balmas, V., Donini, G., Stasse, L., Corazza, L., and Motto, M. (1997). Tolerance to the fungal pathogen *Rhizoctonia solani* AG4 of transgenic tobacco expressing the maize ribosome-inactivating protein b-32. *Transgenic Research* 6: 393-402.

Maher, E.A., Bate, N.J., Ni, W., Elkind, Y., Dixon, R.A., and Lamb, C. (1994). Increased disease susceptibility of transgenic tobacco plants with suppressed levels of preformed phenylpropanoid products. *Proceedings of the National Academy of Sciences, USA* 91: 7802-7806.

Malehorn, D.E., Borgmeyer, J.R., Smith, C.E., and Shah, D.M. (1994). Characterization and expression of an antifungal zeamatin-like protein *(Zlp)* gene from *Zea mays. Plant Physiology* 106: 1471-1481.

Marchant, R., Davey, M.R., Lucas, J.A., Lamb, C.J., Dixon, R.A., and Power, J.B. (1998). Expression of a chitinase transgene in rose (*Rosa hybrida* L.) reduces development of blackspot disease (*Diplocarpon rosae* Wolf). *Molecular Breeding* 4: 187-194.

Masoud, S.A., Zhu, Q., Lamb, C., and Dixon, R.A. (1996). Constitutive expression of an inducible β-1,3-glucanase in alfalfa reduces disease severity caused by the oomycete pathogen *Phytophthora megasperma* f. sp. *medicaginis,* but does not reduce severity of chitin-containing fungi. *Transgenic Research* 5: 313-323.

Mauch, F. and Staehelin, L.A. (1989). Functional implications of the subcellular localization of ethylene-induced chitinase and β-1,3-glucanase in bean leaves. *Plant Cell* 1: 447-457.

Mayer, A.M., Staples, R.C., and Gil-ad, N.L. (2001). Mechanisms of survival of necrotrophic fungal plant pathogens in hosts expressing the hypersensitive response. *Phytochemistry* 58: 33-41.

McDowell, J.M. and Dangl, J.L. (2000). Signal transduction in the plant immune response. *Trends in Biochemical Science* 25: 79-82.

Mehdy, M.C. (1994). Active oxygen species in plant defense against pathogens. *Plant Physiology* 105: 467-472.

Melchers, L.S., Sela-Buurlage, M.B., Vloemans, S.A., Woloshuk, C.P., Van Roekel, J.S.C., Pen, J., Van den Elzen, P.J.M., and Cornelissen, B.J.C. (1993). Extracellular targeting of the vacuolar tobacco proteins AP24, chitinase and ß-1,3-glucanase in transgenic plants. *Plant Molecular Biology* 21: 583-593.

Melchers, L.S. and Stuiver, M.H. (2000). Novel genes for disease-resistance breeding. *Current Opinion in Plant Biology* 3: 147-152.

Métraux, J.P. (2002). Recent breakthroughs in the study of salicylic and biosynthesis. *Trends in Plant Science* 7: 332-334.

Mitsuhara, I., Matsufuru, H., Ohshima, M., Kaku, H., Nakajima, Y., Murai, N., Natori, S., and Ohashi, Y. (2000). Induced expressed of sarcotoxin IA enhanced host resistance against both bacterial and fungal pathogens in transgenic tobacco. *Molecular Plant-Microbe Interactions* 13: 860-868.

Mittler, R., Shulaev, V., and Lam, E. (1995). Coordinated activation of programmed cell death and defense mechanisms in transgenic tobacco plants expressing a bacterial proton pump. *Plant Cell* 7: 29-42.

Mohamed, R., Meilan, R., Ostry, M.E., Michler, C.H., and Strauss, S.H. (2001). Bacterio-opsin gene overexpression fails to elevate fungal disease resistance in transgenic poplar *(Populus). Canadian Journal of Forest Research* 31: 268-275.

Mora, A.A. and Earle, E.D. (2001). Resistance to *Alternaria brassicicola* in transgenic broccoli expressing a *Trichoderma harzianum* endochitinase gene. *Molecular Breeding* 8: 1-9.

Muhitch, M.J., McCormick, S.P., Alexander, N.J., and Hohn, T.M. (2000). Transgenic expression of the *TRI 101* or *PDR 5* gene increases resistance of tobacco to the phytotoxic effects of the trichothecene 4,15-diacetoxyscirpenol. *Plant Science* 157: 201-207.

Murray, F., Llewellyn, D., McFadden, H., Last, D., Dennis, E.S., and Peacock, W.J. (1999). Expression of the *Talaromyces flavus* glucose oxidase gene in cotton and tobacco reduces fungal infection, but is also phytotoxic. *Molecular Breeding* 5: 219-232.

Nakajima, H., Muranaka, T., Ishige, F., Akatsu, K., and Oeda, K. (1997). Fungal and bacterial disease resistance in transgenic plants expressing human lysozyme. *Plant Cell Reports* 16: 674-679.

Neill, S.J., Desikan, R., Clartce, A., Hurst, R.D., and Hancock, J.T. (2002). Hydrogen peroxide and nitric oxide as signaling molecules in plants. *Journal of Experimental Botany* 53: 1237-1247.

Neuhaus, J.-M. (1999). Plant chitinases (PR-3, PR-4, PR-8, PR-11). In Datta S.K., and Muthukrishnan S. (Eds.), *Pathogenesis-Related Proteins in Plants* (pp. 77-105). New York: CRC Press.

Neuhaus, J.-M., Ahl-Goy, P., Hinz, U., Flores, S., and Meins, F. Jr. (1991). High-level expression of a tobacco chitinase gene in *Nicotiana sylvestris:* Susceptibility of transgenic plants to *Cercospora nicotianae* infection. *Plant Molecular Biology* 16: 141-151.

Neuhaus, K., Grsic-Rausch, S., Sauerteig, S., and Ludwig-Müller, J. (2000). *Arabidopsis* plants transformed with nitrilase 1 or 2 in antisense direction are delayed in clubroot development. *Journal of Plant Physiology* 156: 756-761.

Newell, C.A. (2000). Plant transformation technology: Developments and applications. *Molecular Biotechnology* 16: 53-65.

Nicholson, R.L. and Hammerschmidt, R. (1992). Phenolic compounds and their role in disease resistance. *Annual Review of Phytopathology* 30: 369-389.

Nielsen, K.K., Mikkelsen, J.D., Kragh, K.M., and Bojsen, K. (1993). An acidic class III chitinase in sugar beet: Induction by *Cercospora beticola,* characterization, and expression in transgenic tobacco plants. *Molecular Plant-Microbe Interactions* 6: 495-506.

Nishizawa, Y., Nishio, Z., Nakazono, K., Soma, M., Nakajima, E., Ugaki, M., and Hibi, T. (1999). Enhanced resistance to blast *(Magnaporthe grisea)* in transgenic Japonica rice by constitutive expression of rice chitinase. *Theoretical and Applied Genetics* 99: 383-390.

Nuutila, A.M., Ritala, A., Skadsen, R.W., Mannonen, L., and Kauppinen, V. (1999). Expression of fungal thermotolerant endo-1,4-β-glucanase in transgenic barley seeds during germination. *Plant Molecular Biology* 41: 777-783.

O'Donnell, P.J., Calvert, C., Atzorn, R., Wasternack, C., Leyser, H.M.O., and Bowles, D.J. (1996). Ethylene as a signal mediating the wound response of tomato plants. *Science* 274: 1914-1917.

Oldach, K.H., Becker, D., and Lorz, H. (2001). Heterologous expression of genes mediating enhanced fungal resistance in transgenic wheat. *Molecular Plant-Microbe Interactions* 14: 832-838.

Osusky, M., Zhou, G., Osuska, L., Hancock, R.E., Kay, W.W., and Misra, S. (2000). Transgenic plants expressing cationic peptide chimeras exhibit broad-spectrum resistance to phytopathogens. *Nature Biotechnology* 18: 1162-1166.

Pappinen, A., Degefu, Y., Syrjala, L., Keinonen, K., and von Weissenberg, K. (2002). Transgenic silver birch *(Betula pendula)* expressing sugarbeet chitinase 4 shows enhanced resistance to *Pyrenopeziza betulicola*. *Plant Cell Reports* 20: 1046-1051.

Parashina, E.V., Serdobinksii, L.A., Kalle, E.G., Lavrova, N.V., Avetisov, V.A., Lunin, V.G., and Naroditskii, N.V. (2000). Genetic engineering of oilseed rape and tomato plants expressing a radish defensin gene. *Russian Journal of Plant Physiology* 47: 417-423.

Park, C.-M., Berry, J.O., and Bruenn, J.A. (1996). High-level secretion of a virally encoded anti-fungal toxin in transgenic tobacco plants. *Plant Molecular Biology* 30: 359-366.

Powell, A.L.T., Stotz, H.U., Labavitch, J.M., and Bennett, A.B. (1994). Glyco-protein inhibitors of fungal polygalacturonases. In Daniels, M.J., Downie, J.A., and Osbourn, A.E (Eds.), *Advances in Molecular Genetics of Plant-Microbe Interactions*, Volume 3 (pp. 399-402). Dordrecht, the Netherlands: Kluwer Academic Publishers.

Powell, A.L.T., Van Kan, J., ten Have, A., Visser, J., Greve, L.C., Bennett, A.B., and Labavitch, J.M. (2000). Transgenic expression of pear PGIP in tomato limits fungal colonization. *Molecular Plant-Microbe Interactions* 13: 942-950.

Punja, Z.K. (2001). Genetic engineering of plants to enhance resistance to fungal pathogens: A review of progress and future prospects. *Canadian Journal of Plant Pathology* 23: 216-235.

Punja, Z.K. and Raharjo, S.H.T. (1996). Response of transgenic cucumber and carrot plants expressing different chitinase enzymes to inoculation with fungal pathogens. *Plant Disease* 80: 999-1005.

Punja, Z.K. and Zhang, Y.-Y. (1993). Plant chitinases and their roles in resistance to fungal diseases. *Journal of Nematology* 25: 526-540.

Rajasekaran, K., Cary, J.W., Jacks, T.J., Stromberg, K.D., and Cleveland, T.E. (2000). Inhibition of fungal growth in planta and in vitro by transgenic tobacco expressing a bacterial nonheme chloroperoxidase gene. *Plant Cell Reports* 19: 333-338.

Rao, A.G. (1995). Antimicrobial peptides. *Molecular Plant-Microbe Interactions* 8: 6-13.

Ray, H., Douches, D.S., and Hammerschmidt, R. (1998). Transformation of potato with cucumber peroxidase: Expression and disease response. *Physiological and Molecular Plant Pathology* 53: 93-103.

Reymond, P. and Farmer, E.E. (1998). Jasmonate and salicylate as global signals for defense gene expression. *Current Opinions in Plant Biology* 1: 404-411.

Robison, M.M., Shah, S., Tamot, B., Pauls, K.P., Moffatt, B.A., and Glick, B.R. (2001). Reduced symptoms of *Verticillium* wilt in transgenic tomato expressing a bacterial Acc deaminase. *Molecular Plant Pathology* 2: 135-145.

Roby, D., Broglie, K., Cressman, R., Biddle, P., Chet, I., and Broglie, R. (1990). Activation of a bean chitinase promoter in transgenic tobacco plants by phytopathogenic fungi. *Plant Cell* 2: 999-1007.

Rommens, C.M. and Kishore, G.M. (2000). Exploiting the full potential of disease-resistance genes for agricultural use. *Currunt Opinion in Biotechnology* 11: 120-125.

Ryals, J.A., Neuenschwander, U.H., Willits, M.G., Molina, A., Steiner, H.-Y., and Hunt, M.D. (1996). Systemic acquired resistance. *Plant Cell* 8: 1809-1819.

Salmeron, J.M. and Vernooij, B. (1998). Transgenic approaches to microbial disease resistance in crop plants. *Current Opinion in Plant Biology* 1: 347-352.

Schaffrath, U., Mauch, F., Freydl, E., Schweizer, P., and Dudler, R. (2000). Constitutive expression of the defense-related *Rir1b* gene in transgenic rice plants confers enhanced resistance to the rice blast fungus *Magnaporthe grisea*. *Plant Molecular Biology* 43: 59-66.

Schillberg, S., Zimmerman, S., Zhang, M.-Y., and Fischer, R. (2001). Antibody-based resistance to plant pathogens. *Transgenic Research* 10: 1-12.

Schweizer, P., Christoffel, A., and Dudler, R. (1999). Transient expression of members of the *germin*-like gene family in epidermal cells of wheat confers disease resistance. *Plant Journal* 20: 541-552.

Sela-Buurlage, M.B., Ponstein, A.S., Bres-Vloemans, S.A., Melchers, L.S., van den Elzen, P.M.J., and Cornelissen, B.J.C. (1993). Only specific tobacco *(Nicotiana tabacum)* chitinases and β-1,3-glucanases exhibit antifungal activity. *Plant Physiology* 101: 857-863.

Shah, D.M. (1997). Genetic engineering for fungal and bacterial diseases. *Current Opinion in Biotechnology* 8: 208-214.

Shah, D.M., Rommens, C.M.T., and Beachy, R.N. (1995). Resistance to diseases and insects in transgenic plants: Progress and applications to agriculture. *Trends in Biotechnology* 13: 362-368.

Shi, J., Thomas, C.J., King, L.A., Hawes, C.R., Posee, R.D., Edwards, M.L., Pallett, D., and Cooper, J.I. (2000). The expression of a baculovirus-derived chitinase gene increased resistance of tobacco cultivars to brown spot *(Alternaria alternata)*. *Annals of Applied Biology* 136: 1-8.

Shin, R., Park, J.M., An, J.-M, and Paek, K.-H. (2002). Ectopic expression of *Tsi1* in transgenic hot pepper plants enhances host resistance to viral, bacterial, and oomycete pathogens. *Molecular Plant-Microbe Interactions* 15: 983-989.

Shirasu, K. and Schulze-Lefert, P. (2000). Regulators of cell death in disease resistance. *Plant Molecular Biology* 44: 371-385.

Sitbon, F., Hennion, S., Little, C.H.A., and Sundberg, B. (1999). Enhanced ethylene production and peroxidase activity in IAA-overproducing transgenic tobacco plants is associated with increased lignin content and altered lignin composition. *Plant Science* 141: 165-173.

Stark-Lorenzen, P., Nelke, B., Hänbler, G., Mühlbach, H.P., and Thomzik, J.E. (1997). Transfer of a grapevine stillbene synthase gene to rice (*Oryza sativa* L.). *Plant Cell Reports* 16: 668-673.

Staskawicz, B.J., Ausubel, F.M., Baker, B.J., Ellis, J.G., and Jones, J.D.G. (1995). Molecular genetics of plant disease resistance. *Science* 268: 661-667.

Staswick, P.E., Yuen, G.Y., and Lehman, C.C. (1998). Jasmonate signaling mutants of *Arabidopsis* are susceptible to the soil fungus *Pythium irregulare. Plant Journal* 15: 747-754.

Stichter, L., Mauch-Mani, B.N., and Métraux, J.P. (1997). Systemic acquired resistance. *Annual Review of Phytopathology* 35: 235-270.

Stirpe, F., Barbieri, L., Battelli, L.G., Soria, M., and Lappi, D.A. (1992). Ribosome-inactivating proteins from plants: Present status and future prospects. *Bio/Technology* 10: 405-412.

Strittmatter, G., Janssens, J., Opsomer, C., and Botterman, J. (1995). Inhibition of fungal disease development in plants by engineering controlled cell death. *Bio/Technology* 13: 1085-1089.

Strittmatter, G. and Wegner, D. (1993). Genetic engineering of disease and pest resistance in plants: Present state of the art. *Zeischrieft Naturforsch* 480: 673-688.

Swords, K.M.M., Liang, J., and Shah, D.M. (1997). Novel approaches to engineering disease resistance in crops. In Setlow, J.K. (Ed.), *Genetic Engineering,* Volume 19 (pp. 1-13). New York: Plenum Press.

Tabaeizadeh, Z., Agharbaoui, Z., and Harrak, H. (1999). Transgenic tomato plants expressing a *Lycopersicon chilense* chitinase gene demonstrate improved resistance to *Verticillium dahliae* race 2. *Plant Cell Reports* 19: 197-202.

Tabei, Y., Kitade, S., Nishizawa, Y., Kikuchi, N., Kayano, T., Hibi, T., and Akutsu, K. (1998). Transgenic cucumber plants harboring a rice chitinase gene exhibit enhanced resistance to gray mold *(Botrytis cinerea). Plant Cell Reports* 17: 159-164.

Takahashi, H., Chen, Z., Du, H., Liu, Y., and Klessig, D.F. (1997). Development of necrosis and activation of disease resistance in transgenic tobacco plants with severely reduced catalase levels. *Plant Journal* 11: 993-1005.

Takaichi, M. and Oeda, K. (2000). Transgenic carrots with enhanced resistance against two major pathogens, *Erysiphe heraclei* and *Alternaria dauci. Plant Science* 153: 135-144.

Takatsu, Y., Nishizawa, Y., Hibi, T., and Akutsu, K. (1999). Transgenic chrysanthemum [*Dendranthema grandiflorum* (Ramat.) Kitamura] expressing a rice chitinase gene shows enhanced resistance to gray mold *(Botrytis cinerea). Scientia Horticulturae* 82: 113-123.

Takken, F.L.W. and Joosten, M.H.A.J. (2000). Plant resistance genes: Their structure, function and evolution. *European Journal of Plant Pathology* 106: 699-713.

Tenhaken, R., Levine, A., Brisson, L.F., Dixon, R.A., and Lamb, C. (1995). Function of the oxidative burst in hypersensitive disease resistance. *Proceedings of the National Academy of Sciences, USA* 92: 4158-4163.

Tepfer, D., Boutteaux, C., Vigon, C., Aymes, S., Perez, V., O'Donohue, M.J., Huet, J.-C., and Pernollet, J.-C. (1998). *Phytophthora* resistance through production of a fungal protein elicitor (β-cryptogein) in tobacco. *Molecular Plant-Microbe Interactions* 11: 64-67.

Terakawa, T., Takaya, N., Horiuchi, H., Koike, M., and Takagi, M. (1997). A fungal chitinase gene from *Rhizopus oligosporus* confers antifungal activity to transgenic tobacco. *Plant Cell Reports* 16: 439-443.

Terras, F.R.G., Eggermont, K., Kovaleva, V., Raikhel, N.V., Osborn, R.W., Kester, A., Rees, S.B., Torrekens, S., Van Leuven, F., Vanderleyden, J., Cammue, B.PA., and Broekaert, W.F. (1995). Small cystein-rich antifungal proteins from radish: Their role in host defence. *Plant Cell* 7: 573-588.

Thomas, C.M., Jones, D.A., Parniske, M., Harrison, K., Balint-Kurti, P.J., Hatzixanthis, K., and Jones, J.D.G. (1997). Characterization of the tomato *Cf-4* gene for resistance to *Cladosporium fulvum* identifies sequences that determine recognitional specificity in Cf-4 and Cf-9. *Plant Cell* 9: 2209-2224.

Thomma, B.P.H.J., Eggermont, K., Tierens, K.F.M.-J., and Broekaert, W.F. (1999). Requirement of functional *ethylene-insensitive* 2 gene for efficient resistance of *Arabidopsis* to infection by *Botrytis cinerea*. *Plant Physiology* 121: 1093-1101.

Thomma, B.P.H.J., Nelissen, I., Eggermont, K., and Broekaert, W.F. (1999). Deficiency in phytoalexin production causes enhanced susceptibility of *Arabidopsis thaliana* to the fungus *Alternaria brassicicola*. *Plant Journal* 19: 163-171.

Thompson, C., Dunwell, J.M., Johnstone, C.E., Lay, V., Ray, J., Schmitt, M., Watson, H., and Nisbet, G. (1995). Degradation of oxalic acid by transgenic oilseed rape plants expressing oxalate oxidase. *Euphytica* 85: 169-172.

Thomzik, J.E., Stenzel, K., Stöcker, R., Schreier, P.H., Hain, R., and Stahl, D.J. (1997). Synthesis of a grapevine phytoalexin in transgenic tomatoes (*Lycopersicon esculentum* Mill.) conditions resistance against *Phytophthora infestans*. *Physiological and Molecular Plant Pathology* 51: 265-278.

Uchimiya, H., Fujii, S., Huang, J., Fushimi, T., Nishioka, N., Kim, K.-M., Yamada, M.K., Kurusu, T., Kuchitsu, K., and Tagawa, M. (2002). Transgenic rice plants conferring increased tolerance to rice blast and multiple environtal stresses. *Molecular Breeding* 9: 25-31.

Van den Elzen, P.J.M., Jongedijk, E., Melchers, L.S., and Cornelissen, B.J.C. (1993). Virus and fungal resistance: From laboratory to field. *Philosophical Transactions of the Royal Society of London* B 342: 271-278.

Van Loon, L.C. and Van Strien, E.A. (1999). The families of pathogenesis-related proteins, their activities, and comparative analysis of PR-1 type proteins. *Physiological and Molecular Plant Pathology* 55: 85-97.

Verberne, M.C., Verpoorte, R., Bol, J.F., Mercado-Blanco, J., and Linthorst, H.J.M. (2000). Overproduction of salicyclic acid in plants by bacterial transgenes enhances pathogen resistance. *Nature Biotechnology* 18: 779-783.

Vierheilig, H., Alt, M., Lange, J., Gut-Rella, M., Wiemken, A., and Boller, T. (1995). Colonization of transgenic tobacco constitutively expressing pathogenesis-related proteins by the vesicular-arbuscular mycorrhizal fungus *Glomus mosseae*. *Applied and Environmental Microbiology* 61: 3031-3034.

Vierheilig, H., Alt, M., Neuhaus, J.-M., Boller, T., and Wiemken, A. (1993). Colonization of transgenic *Nicotiana sylvestris* plants, expressing different forms of

Nicotiana tabacum chitinase, by the root pathogen *Rhizoctonia solani* and by the mycorrhizal symbiont *Glomus mosseae*. *Molecular Plant-Microbe Interactions* 6: 261-264.

Vijayan, P., Shockey, J., Lévesque, C.A., Cook, R.J., and Browse, J. (1998). A role for jasmonate in pathogen defense of *Arabidopsis*. *Proceedings of the National Academy of Sciences, USA* 95: 7209-7214.

Walton, J.D. (1994). Deconstructing the cell wall. *Plant Physiology* 104: 1113-1118.

Wang, P., Zoubenko, O., and Tumer, N.E. (1998). Reduced toxicity and broad spectrum resistance to viral and fungal infection in transgenic plants expressing pokeweed antiviral protein II. *Plant Molecular Biology* 38: 957-964.

Wang, Y., Nowak, G., Culley, D., Hadwiger, L.A., and Fristensky, B. (1999). Constitutive expression of a pea defense gene DRR206 confers resistance to blackleg *(Leptosphaeria maculans)* disease in transgenic canola *(Brassica napus)*. *Molecular Plant-Microbe Interactions* 12: 410-418.

Wattad, C., Kobiler, D., Dinoor, A., and Prusky, D. (1997). Pectate lyase of *Colletotrichum gloeosporioides* attacking avocado fruits—cDNA cloning and involvement in pathogenicity. *Physiological Plant Pathology* 50: 197-212.

Weiler, E.W. (1997). Octadecanoid-mediated signal transduction in higher plants. *Naturwissenschaften* 84: 340-349.

Wilkinson, J.Q., Lanahan, M.B., Clark, D.G., Bleecker, A.B., Chang, C., Meyerowitz, E.M., and Klee, H.J. (1997). A dominant mutant receptor from *Arabidopsis* confers ethylene insensitivity in heterologous plants. *Nature Biotechnology* 15: 444-447.

Woloshuk, C.P., Meulenhoff, J.S., Sela-Buurlage, M., van den Elzen, P.J.M., and Cornelissen, B.J.C. (1991). Pathogen-induced proteins with inhibitory activity toward *Phytophthora infestans*. *Plant Cell* 3: 619-628.

Wong, K.W., Harman, G.E., Norelli, J.L., Gustafson, H.L., and Aldwinckle, H.S. (1999). Chitinase-transgenic lines of 'Royal Gala' apple showing enhanced resistance to apple scab. *Acta Horticulturae* 484: 595-599.

Wu, G., Shortt, B.J., Lawrence, E.B., Léon, J., Fitzsimmons, K.C., Levine, E.B., Raskin, I., and Shah, D.M. (1997). Activation of host defense mechanisms by elevated production of H_2O_2 in transgenic plants. *Plant Physiology* 115: 427-435.

Wu, G., Shortt, B.J., Lawrence, E.B., Levine, E.B., Fitzsimmons, K.C., and Shah, D.M. (1995). Disease resistance conferred by expression of a gene encoding H_2O_2-generating glucose oxidase in transgenic potato plants. *Plant Cell* 7: 1357-1368.

Yamamoto, T., Iketani, H., Ieki, H., Nishizawa, Y., Notsuka, K., Hibi, T., Hayashi, T., and Matsuta, N. (2000). Transgenic grapevine plants expressing a rice chitinase with enhanced resistance to fungal pathogens. *Plant Cell Reports* 19: 639-646.

Yang, Y., Shah, J., and Klessig, D.F. (1997). Signal perception and transduction in plant defense responses. *Genes and Development* 11: 1621-1639.

Yao, K., De Luca, V., and Brisson, N. (1995). Creation of a metabolic sink for tryptophan alters the phenylpropanoid pathway and the susceptibility of potato to *Phytophthora infestans*. *Plant Cell* 7: 1787-1799.

Yoshikawa, M., Tsuda, M., and Takeuchi, Y. (1993). Resistance to fungal diseases in transgenic tobacco plants expressing the phytoalexin elicitor-releasing factor, β-1,3-endoglucanase, from soybean. *Naturwissenschaften* 80: 417-420.

Yu, D., Liu, Y., Fan, B., Klessig, D.F., and Chen, Z. (1997). Is the high basal level of salicyclic acid important for disease resistance in potato? *Plant Physiology* 115: 343-349.

Yu, D., Xie, Z., Chen, C., Fan, B., and Chen, Z. (1999). Expression of tobacco class II catalase gene activates the endogenous homologous gene and is associated with disease resistance in transgenic potato plants. *Plant Molecular Biology* 39: 477-488.

Yuan, H., Ming, X., Wang, L., Hu, P., An, C., and Chen, Z. (2002). Expression of a gene encoding trichosanthin in transgenic rice plants enhances resistance to fungus blast disease. *Plant Cell Reports* 20: 992-998.

Yun, D.-J., Bressan, R.A., and Hasegawa, P.M. (1997). Plant antifungal proteins. In Janick, J. (Ed.). *Plant Breeding Reviews*, Volume 14. (pp. 39-88). New York: John Wiley and Sons.

Zhang, Z., Collinge, D.B., and Thordal-Christensen, H. (1995). Germin-like oxalate oxidase, a H_2O_2-producing enzyme, accumulates in barley attacked by the powdery mildew fungus. *Plant Journal* 8: 139-145.

Zhu, B., Chen, T.H.H., and Li, P.H. (1995). Activation of two osmotin-like protein genes by abiotic stimuli and fungal pathogen in transgenic potato plants. *Plant Physiology* 108: 929-937.

Zhu, B., Chen, T.H.H., and Li, P.H. (1996). Analysis of late-blight disease resistance and freezing tolerance in transgenic potato plants expressing sense and antisense genes for an osmotin-like protein. *Planta* 198: 70-77.

Zhu, Q., Maher, E.A., Masoud, S., Dixon, R.A., and Lamb, C.J. (1994). Enhanced protection against fungal attack by constitutive co-expression of chitinase and glucanase genes in transgenic tobacco. *Bio/Technology* 12: 807-812.

Zook, M., Hohn, T., Bonnen, A., Tsuji, J., and Hammerschmidt, R. (1996). Characterization of novel sesquiterpenoid biosynthesis in tobacco expressing a fungal sesquiterpene synthase. *Plant Physiology* 112: 311-318.

Zoubenko, O., Uckun, F., Hur, Y., Chet, I., and Tumer, N. (1997). Plant resistance to fungal infection induced by nontoxic pokeweed antiviral protein mutants. *Nature Biotechnology* 15: 992-996.

TABLE 7.1. Plant species genetically engineered to enhance resistance to fungal diseases (1991-2002)

Strategy used and plant species engineered	Expressed gene product	Effect on disease development	Reference
Expression of hydrolytic enzymes			
Alfalfa (*Medicago sativa* L.)	Alfalfa glucanase	Reduced symptom development due to *Phytophthora megasperma*; no effect on *Stemphylium alfalfae*	Masoud et al., 1996
American ginseng (*Panax quinquefolius* L.)	Rice chitinase	Not tested	Chen and Punja, 2002a
Apple (*Malus ×domestica*)	*Trichoderma harzianum* endochitinase	Reduced lesion number and lesion area due to *Venturia inaequalis*	Bolar et al., 2000; Wong et al., 1999
Barley (*Hordeum vulgare* L.)	*Trichoderma* endo-1,4-β-glucanase	Not tested	Nuutila et al., 1999
Broccoli (*Brassica oleracea* var. *italica*)	*Trichoderma harzianum* endochitinase	Reduced lesion size due to *Alternaria brassicicola*	Mora and Earle, 2001
Canola (*Brassica napus* L.)	Bean chitinase	Reduced rate and total seedling mortality due to *Rhizoctonia solani*	Broglie et al., 1991

Plant	Chitinase source	Effect	Reference
Canola (*B. napus* L.)	Tomato chitinase	Lower percentage of diseased plants due to *Cylindrosporium concentricum* and *Sclerotinia sclerotiorum*	Grison et al., 1996
Carrot (*Daucus carota* L.)	Tobacco chitinase	Reduced rate and final incidence of disease due to *Botrytis cinerea*, *Rhizoctonia solani*, and *Sclerotium rolfsii*; no effect on *Thielaviopsis basicola* and *Alternaria radicina*	Punja and Raharjo, 1996
Chrysanthemum [*Dendranthema grandiflorum* (Ramat.) Kitamura]	Rice chitinase	Reduced lesion development due to *Botrytis cinerea*	Takatsu et al., 1999
Creeping bentgrass (*Agrostis palustris* Huds.)	Elm chitinase-like protein	Reduced severity of *Rizoctonia solani*	Chai et al., 2002
Cucumber (*Cucumis sativus* L.)	Petunia and tobacco chitinases	No effect on disease development due to *Colletotrichum lagenarium* and *Rhizoctonia solani*	Punja and Raharjo, 1996
Cucumber (*C. sativus* L.)	Rice chitinase	Reduced lesion development due to *Botrytis cinerea*	Tabei et al., 1998
Grape (*Vitis vinifera* L.)	Rice chitinase	Reduced development of *Uncinula necator* and fewer lesions due to *Elsinoe ampelina*	Yamamoto et al., 2000

TABLE 7.1 (continued)

Plant	Gene	Effect	Reference
Grape (*V. vinifera* L.)	*Trichoderma harzianum* endochitinase	Reduction of *Botrytis cinerea* development in preliminary tests	Kikkert et al., 2000
Potato (*Solanum tuberosum* L.)	*Trichoderma harzianum* endochitinase	Lower lesion numbers and size due to *Alternaria solani*; reduced mortality due to *Rhizoctonia solani*	Lorito et al., 1998
Rice (*Oryza sativa* L.)	Rice chitinase	Fewer numbers of lesions and smaller size due to *Rhizoctonia solani*	Lin et al., 1995; Datta et al., 2000, 2001
Rice (*O. sativa* L.)	Rice chitinase	Delayed onset and reduced severity of disease symptoms due to *Magnaporthe grisea*	Nishizawa et al., 1999
Rose (*Rosa hybrida* L.)	Rice chitinase	Reduced lesion diameter due to black spot (*Diplocarpon rosae*)	Marchant et al., 1998
Silver birch (*Betula pendula* L.)	Sugarbeet chitinase	Enhanced resistance to *Pyrenopeziza betulicola*	Pappinen et al., 2002
Sorghum [*Sorghum bicolor* (L.) Moench]	Rice chitinase	Increased resistance to *Fusarium thapsinum*	Krishnaveni et al., 2001
Strawberry (*Fragaria xananassa* Duch.)	Rice chitinase	Reduced development of powdery mildew (*Sphaerotheca humuli*)	Asao et al., 1997

Plant	Gene/Enzyme	Effect	Reference
Tobacco (*Nicotinana tabacum* L.)	Bean chitinase	Lower seedling mortality due to *Rhizoctonia solani*; no effect on *Pythium aphanidermatum*	Broglie et al., 1991, Broglie et al., 1993
Tobacco (*N. tabacum* L.)	Peanut chitinase	Not tested	Kellmann et al., 1996
Tobacco (*N. tabacum* L.)	*Serratia marcescens* chitinase	Reduced disease incidence due to *Rhizoctonia solani* on seedlings; no effect on *Pythium ultimum*	Howie et al., 1994
Tobacco (*N. tabacum* L.)	*Serratia marcescens* chitinase	Reduced development of *Rhizoctonia solani*	Jach et al., 1992
Tobacco (*N. tabacum* L.)	*Streptomyces* chitosanase	Not tested	El Quakfaoui et al., 1995
Tobacco (*N. tabacum* L.)	*Rhizopus oligosporus* chitinase	Reduced rate of development and size of lesions on leaves due to *Botrytis cinerea* and *Sclerotinia sclerotiorum*	Terakawa et al., 1997
Tobacco (*N. tabacum* L.)	*Trichoderma harzianum* endochitinase	Reduced symptoms due to *Alternaria alternata*, *Botrytis cinerea*, and *Rhizoctonia solani*	Lorito et al., 1998
Tobacco (*N. tabacum* L.)	Baculovirus chitinase	Reduced lesion development due to brown spot (*Alternaria alternata*)	Shi et al., 2000

TABLE 7.1 (continued)

Tobacco (*N. tabacum* L.)	Soybean glucanase	Reduced development of *Phytophthora parasitica* and *Alternaria alternata*	Yoshikawa et al., 1993
Tobacco (*N. tabacum* L.)	Tobacco glucanase	Reduced disease symptoms due to *Phytophthora parasitica* and *Peronospora tabacina*	Lusso and Kuc, 1996
Tobacco (*N. tabacum* L.)	*Acidothermus cellulolyticus* endoglucanase	Not tested	Dai et al., 2000
Tobacco (*N. benthamiana* L.)	Sugarbeet chitinase	No effect on *Cercospora nicotianae*	Nielsen et al., 1993
Tobacco (*N. sylvestris* L.)	Tobacco chitinase	No effect on *Cercospora nicotianae*	Neuhaus et al., 1991
Tobacco (*N. sylvestris* L.)	Tobacco chitinase	Reduced colonization by *Rhizoctonia solani*	Vierheilig et al., 1993
Tomato (*Lycopersicon esculentum* Mill.)	Wild tomato (*L. chilense*) chitinase	Reduced development of *Verticillium dahliae* races 1 and 2	Tabaeizadeh et al., 1999
Wheat (*Triticum aestivum* L.)	Barley chitinase	Reduced development of colonies of *Blumeria graminis* f. sp. *tritici*	Bliffeld et al., 1999

Plant	Gene/protein	Effect	Reference
Wheat (*T. aestivum* L.)	Barley chitinase	Reduced development of colonies of *Blumeria graminis* f. sp. *tritici* and *Puccinia recondita* f. sp. *tritici*	Oldach et al., 2001

Expression of pathogenesis-related (PR) proteins

Plant	Gene/protein	Effect	Reference
Canola (*Brassica napus* L.)	Pea chitinase, PR10.1 gene	No effect on *Leptosphaeria maculans*	Wang et al., 1999
Canola (*B. napus* L.)	Pea defense response gene, defensin	Reduced infection and development of *Leptosphaeria maculans*	Wang et al., 1999
Carrot (*D. carota* L.)	Rice thaumatin-like protein	Reduced rate and final disease incidence due to *Botrytis cinerea* and *Sclerotinia sclerotiorum*	Chen and Punja, 2002b
Potato (*S. tuberosum* L.)	Tobacco osmotin	Delayed onset and rate of disease due to *Phytophthora infestans*	Liu et al., 1994
Potato (*S. commersonii* Dun.)	Potato osmotin-like protein	Enhanced tolerance to infection by *Phytophthora infestans*	Zhu et al., 1996
Potato (*S. tuberosum* L.)	Pea PR10 gene	Reduced development of *Vert-icillium dahliae*	Chang et al., 1993
Potato (*S. tuberosum* L.)	Potato defense response gene STH-2	No effect against *Phytophthora infestans*	Constabel et al., 1993

TABLE 7.1 (continued)

Rice (*O. sativa* L.)	Rice thaumatin-like protein	Reduced lesion development due to *Rhizoctonia solani*	Datta et al., 1999
Rice (*O. sativa* L.)	Rice Rir1b defense gene	Fewer lesions due to *Magnaporthe grisea*	Schaffrath et al., 2000
Tobacco (*N. tabacum* L.)	Tobacco PR1a	Reduced rate and final disease due to *Peronospora tabacina* and *Phytophthora parasitica*	Alexander et al., 1993
Tobacco (*N. tabacum* L.)	Tobacco osmotin	No effect on *Phytophthora parasitica* var. *nicotianae*	Liu et al., 1994
Wheat (*T. aestivum* L.)	Rice thaumatin-like protein	Delayed development of *Fusarium graminearum*	Chen et al., 1999

Expression of antimicrobial proteins/peptides/compounds

Arabidopsis (*A. thaliana* L.)	Mistletoe thionin viscotoxin	Reduced infection and development of *Plasmodiophora brassicae*	Holtorf et al., 1998
Arabidopsis (*A. thaliana* L.)	*Arabidopsis* thionin	Reduced development and colonization by *Fusarium oxysporum*	Epple et al., 1997
Canola (*B. napus* L.)	Macadamia antimicrobial peptide	Reduced lesion size due to *Leptosphaeria maculans*	Kazan et al., 2002
Carrot (*Daucus carota* L.)	Human lysozyme	Enhanced resistance to *Erysiphe heraclei* and *Alternaria dauci*	Takaichi and Oeda, 2000

Plant	Protein	Effect	Reference
Geranium (*Pelargonium* sp.)	Onion antimicrobial protein	Reduced development and sporulation of *Botrytis cinerea*	Bi et al., 1999
Indian mustard (*Brassica juncea* L.)	*Hevea* chitin-binding lectin (hevein)	Smaller lesion size and reduced rate of development due to *Alternaria brassicae*	Kanrar et al., 2002
Poplar (*Populus ×euramericana*)	Antimicrobial peptide	Reduced lesion size due to *Septoria musiva*	Liang et al., 2002
Potato (*S. tuberosum* L.)	Alfalfa defensin	Enhanced resistance to *Verticillium dahliae*	Gao et al., 2000
Potato (*S. tuberosum* L.)	*Bacillus amyloliquefaciens* barnase (RNase)	Delayed sporulation and reduced sporangia production by *Phytophthora infestans*	Strittmatter et al., 1995
Potato (*S. tuberosum* L.)	Synthetic cationic peptide chimera	Reduced development of *Fusarium solani* and *Phytophthora cactorum*	Osusky et al., 2000
Potato (*S. tuberosum* L.)	Human lactoferrin	Not tested	Chong and Langridge, 2000
Rice (*O. sativa*)	Maize ribosome-inactivating protein	No effect on *Magnaporthe grisea* or *R. solani*	Kim et al., 1999
Rice (*O. sativa* L.)	*Trichosanthes* ribosome-inactivating protein	Reduced lesion size due to *Pyricularia oryzae* and enhanced seedling survival	Yuan et al., 2002

TABLE 7.1 (continued)

Rice (*O. sativa* L.)	Wheat puroindoline peptide	Reduced symptoms due to *Magnaporthe grisea* and *Rhizoctonia solani*	Krishnamurthy et al., 2001
Tobacco (*N. tabacum* L.)	*Amaranthus* hevein-type peptide, *Mirabilis* knottin-type peptide	No effect on *Alternaria longipes* or *Botrytis cinerea*	De Bolle et al., 1996
Tobacco (*N. tabacum* L.)	Radish defensin	Reduced infection and lesion size due to *Alternaria longipes*	Terras et al., 1995
Tobacco (*N. tabacum* L.)	Stinging nettle (*Urtica dioica* L.) isolectin	Not tested	Does et al., 1999
Tobacco (*N. tabacum* L.)	Pokeweed antiviral protein	Lower rate of infection and mortality due to *Rhizoctonia solani*	Wang et al., 1998; Zoubenko et al., 1997
Tobacco (*N. tabacum* L.)	Synthetic magainin-type peptide	Reduced lesion development due to *Colletotrichum destructivum* and *Peronospora tabacina*	DeGray et al., 2001; Li et al., 2001
Tobacco (*N. tabacum* L.)	Sarcotoxin peptide from *Sarcophaga peregrina*	Enhanced seedling survival following inoculation with *R. solani*, *Pythium aphanidermatum*, and *Phytophthora nicotianae*	Mitsuhara et al., 2000

Tobacco (N. tabacum L.)	Barley ribosome-inactivating protein	Reduced incidence and severity of Rhizoctonia solani	Logemann et al., 1992
Tobacco (N. tabacum L.)	Maize ribosome-inactivating protein	Lower damage due to Rhizoctonia solani	Maddaloni et al., 1997
Tobacco (N. tabacum L.)	Antifungal (killing) protein from Ustilago maydis infecting virus (dsRNA)	Not tested	Park et al., 1996
Tobacco (N. tabacum L.)	Chloroperoxidase from Pseudomonas pyrrocinia	Reduced lesion development by Colletotrichum destructivum	Rajasekaran et al., 2000
Tobacco (N. tabacum L.)	Synthetic antimicrobial peptide	Reduced lesion size due to Colletotrichum destructivum	Cary et al., 2000
Tobacco (N. tabacum L.)	Human lysozyme	Reduced colony size and conidial production by Erysiphe cichoracearum	Nakajima et al., 1997
Tobacco (N. tabacum L.)	Insect antifungal peptides	Reduced development of Cercospora nicotianae	Banzet et al., 2002

TABLE 7.1 (continued)

Tomato (*L. esculentum* Mill.)	Radish defensin	Reduced number and size of lesions due to *Alternaria solani*	Parashina et al., 2000
Wheat (*T. aestivum* L.)	Barley ribosome-inactivating protein	Slightly reduced development of *Blumeria graminis*	Bieri et al., 2000
Wheat (*T. aestivum* L.)	Antifungal (killing) protein from *Ustilago maydis*-infecting virus (dsRNA)	Inhibition of *Ustilago maydis* and *Tilletia tritici* development on seeds	Clausen et al., 2000
Wheat (*T. aestivum* L.)	*Aspergillus* antifungal protein	Reduced development of colonies of *Blumeria graminis* f. sp. *tritici* and *Puccinia recondita* f. sp. *tritici*	Oldach et al., 2001
Expression of phytoalexins			
Alfalfa (*M. sativa* L.)	Alfalfa isoflavone *O*-methyltransferase	Reduced lesion size due to *Phoma medicaginis*	He and Dixon, 2000
Alfalfa (*M. sativa* L.)	Peanut resveratrol synthase	Reduced lesion size and sporulation of *Phoma medicaginis*	Hipskind and Paiva, 2000
Barley (*H. vulgare*)	Grape stilbene (resveratrol) synthase	Reduced colonization by *Botrytis cinerea*	Leckband and Lörz, 1998

Plant	Transgene product	Effect	Reference
Grape (*V. vinifera* L.)	Grape stilbene (resveratrol) synthase	Reduced colonization by *Botrytis cinerea*	Coutos-Thevenot et al., 2001
Rice (*O. sativa* L.)	Grape stilbene (resveratrol) synthase	Reduced lesion development due to *Pyricularia oryzae*	Stark-Lorenzen et al., 1997
Tobacco (*N. tabacum* L.)	Synthetic magainin-type peptide	Reduced lesion size and sporulation due to *Peronospora tabacina*	Li et al., 2001
Tobacco (*N. tabacum* L.)	*Fusarium* trichodiene synthase	Not tested	Zook et al., 1996
Tobacco (*N. tabacum* L.)	Grape stilbene (resveratrol) synthase	Reduced colonization by *Botrytis cinerea*	Hain et al., 1993
Tomato (*L. esculentum* Mill.)	Grape stilbene (resveratrol) synthase	Reduced lesion development by *Phytophthora infestans*; no effect on *Alternaria solani* or *Botrytis cinerea*	Thomzik et al., 1997
Wheat (*T. aestivum* L.)	Grape stilbene (resveratrol) synthase	Not tested	Fettig and Hess, 1999

Inhibition of pathogen virulence products

Canola (*B. napus* L.)	Barley oxalate oxidase	Not tested	Thompson et al., 1995

TABLE 7.1 (continued)

Poplar (*Populus xeuramericana*)	Wheat oxalate oxidase	Delayed development of *Septoria musiva*	Liang et al., 2001
Rice (*O. sativa* L.)	Rice HC-toxin reductase-like	Reduced lesion development due to *Magnaporthe grisea*	Uchimiya et al., 2002
Soybean [*Glycine max* (L.) Merrill]	Wheat oxalate oxidase (germin)	Reduced lesion length and disease progression due to Sclerotina sclerotiorum	Donaldson et al., 2001
Tobacco (*N. tabacum* L.)	*Fusarium* trichothecene-degrading enzyme	Not tested	Muhitch et al., 2000
Tobacco (*N. tabacum* L.)	Mutant *RpL3* gene for mycotoxin insensitity	Enhanced tolerance to *Fusarium graminearum* mycotoxin	Harris and Gleddie, 2001
Tobacco (*N. tabacum* L.)	Wheat oxalate oxidase (germin)	Not tested	Berna and Bernier, 1997
Tobacco (*N. umbratica* L.)	Tomato *Asc-1* gene for insensitivity to fungal toxins	Enhanced resistance to *Alternaria alternata* f. sp. *lycopersici*	Brandwagt et al., 2002
Tomato (*L. esculentum* Mill.)	Bean polygalacturonase inhibiting protein	No effect on disease due to *Fusarium oxysporum*, *Botrytis cinerea*, and *Alternaria solani*	Desiderio et al., 1997

Tomato (*L. esculentum* Mill.)	Pear polygalacturonase inhibiting protein	Reduced rate of development of *Botrytis cinerea*	Powell et al., 2000
Tomato (*L. esculentum* Mill.)	*Collybia velutipes* oxalate decarboxylase	Enhanced resistance to *Sclerotinia sclerotiorum*	Kesarwani et al., 2000

Alteration of structural components

Potato (*S. tuberosum* L.)	Cucumber peroxidase	No effect on disease due to *Fusarium sambucinum* and *Phytophthora infestans*	Ray et al., 1998
Tomato (*L. esculentum* Mill.)	Tobacco anionic peroxidase	No effect on disease due to *Fusarium oxysporum* and *Verticillium dahliae*	Lagrimini et al., 1993
Wheat (*T. aestivum* L.)	Wheat germin (no oxalate oxidase activity)	Reduced penetration by *Erysiphe blumeria* into epidermal cells	Schweizer et al., 1999

Regulation of plant defense responses

Arabidaopsis (*A. thaliana* L.)	*Arabidopsis* NPR1 protein	Reduced infection and growth of *Peronospora parasitica*	Cao et al., 1998
Arabidaopsis (*A. thaliana* L.)	*Arbidopsis* ethyl-ene-response-factor 1(ERF1)	Enhanced tolerance to *Botrytis cinerea* and *Plectosphaerella cucumerina*	Berrocal-Lobo et al., 2002
Canola (*Brassica napus* L.)	Tomato *Cf9* gene	Delayed disease development due to *Leptosphaeria maculans*	Hennin et al., 2001

TABLE 7.1 (continued)

Cotton (*Gossypium hirsutum* L.), tobacco (*N. tabacum* L.)	*Talaromyces flavus* glucose oxidase	Enhanced protection against *Rhizoctonia solani*, and *Verticillium dahliae*; no effect on *Fusarium oxysporum*	Murray et al., 1999
Pepper (*Capsicum annuum* L.)	Tobacco ethylene-responsive protein	Enhanced resistance to *Phytophthora capsici*	Shin et al., 2002
Potato (*S. tuberosum* L.)	*Aspergillus niger* glucose oxidase	Delayed lesion development due to *Phytophthora infestans*; reduced disease development due to *Alternaria solani* and *Verticil-lium dahliae*	Wu et al., 1995; 1997
Potato (*S. tuberosum* L.)	Tobacco catalase	Reduced lesion size due to *Phytophthora infestans*	Yu et al., 1999
Potato (*S. tuberosum* L.)	Bacterial salicylate hydroxylase	No effect on *Phytophthora infestans*	Yu et al., 1997
Tobacco (*N. tabacum* L.)	*Aspergillus niger* glucose oxidase	Delayed disease development due to *Phytophthora nicotianae*	Lee et al., 2002
Tobacco (*N. tabacum* L.), *Arabidopsis* (*A. thaliana* L.)	Bacterial salicylate hydroxylase	Enhanced susceptibility to *Phytophthora parasitica*, *Cercospora nicotianae*, and *Peronospora parasitica*	Delaney et al., 1994; Donofrio and Delaney, 2001

256

Tobacco (*N. tabacum* L.)	Bacterial salicylic acid-generating enzymes	Enhanced resistance to *Oidium lycopersici*	Verberne et al., 2000
Tobacco (*N. tabacum* L.)	*Arabidopsis* ethylene-insensitivity gene	Enhanced susceptibility to *Pythium sylvaticum*	Knoester et al., 1998
Tobacco (*N. tabacum* L.)	*Phytophthora cryptogea* elicitor (β-cryptogein)	Reduced infection by *Phytophthora parasitica*	Tepfer et al., 1998
Tobacco (*N. tabacum* L.)	*Phytophthora cryptogea* elicitor (cryptogein)	Enhanced resistance to *Phytophthora parasitica, Thielaviopsis basicola, Botrytis cinerea,* and *Erysiphe cichoracearum*	Keller et al., 1999
Tomato (*L. esculentum* Mill.)	*Enterobacter* ACC deaminase	Reduced symptom development due to *Verticillium dahliae*	Robison et al., 2001

Expression of combined gene products

Apple (*Malus ×domestica*)	*Trichoderma atroviride* endochitinase + exochitinase	Increased resistance to *Venturia inaequalis*	Bolar et al., 2001

257

TABLE 7.1 *(continued)*

Carrot (*D. carota* L.)	Tobacco chitinase + β-1-3-glucanase, osmotin	Enhanced resistance to *Alternaria dauci, A. radicina, Cercospora carotae,* and *Erysiphe heraclei*	Melchers and Stuiver, 2000
Tobacco (*N. tabacum* L.)	Barley chitinase + β-1,3-glucanase, or chitinase + ribosome-inactivating protein	Reduced disease severity due to *Rhizoctonia solani*	Jach et al., 1995
Tobacco (*N. tabacum* L.)	Rice chitinase + alfalfa glucanase	Reduced rate of lesion development and fewer lesions due to *Cercospora nicotianae*	Zhu et al., 1994
Tomato (*L. esculentum* Mill.)	Tobacco chitinase + β-1,3-glucanase	Reduced disease severity due to *Fusarium oxysporum* f. sp. *lycopersici*	Jongedijk et al., 1995; van den Elzen et al., 1993

Source: Reproduced by permission from the Canadian Phytopathological Society. Adapted from Punja, 2001.

Index

Page numbers followed by the letter "f" indicate figures; those followed by the letter "t" indicate tables.